Harcourt Math
Georgia Edition

Practice/Homework Workbook

TEACHER EDITION
Grade 2

Harcourt

Visit The Learning Site!
www.harcourtschool.com

Copyright © by Harcourt, Inc.

All rights reserved. No part of this publication may be reproduced or transmitted in any form or by any means, electronic or mechanical, including photocopy, recording, or any information storage and retrieval system, without permission in writing from the publisher.

Requests for permission to make copies of any part of the work should be addressed to School Permissions and Copyrights, Harcourt, Inc., 6277 Sea Harbor Drive, Orlando, Florida 32887-6777. Fax: 407-345-2418.

HARCOURT and the Harcourt Logo are trademarks of Harcourt, Inc., registered in the United States of America and/or other jurisdictions.

Printed in the United States of America

ISBN 13: 978-0-15-349546-5
ISBN 10: 0-15-349546-4

If you have received these materials as examination copies free of charge, Harcourt School Publishers retains title to the materials and they may not be resold. Resale of examination copies is strictly prohibited and is illegal.

Possession of this publication in print format does not entitle users to convert this publication, or any portion of it, into electronic format.

4 5 6 7 8 9 10 1413 15 14 13 12 11 10 09

CONTENTS

Unit 1: NUMBERS AND OPERATIONS

Chapter 1: Numbers to 100
- 1.1 Tens 1
- 1.2 Tens and Ones 2
- 1.3 Understand Place Value 3
- 1.4 Read and Write Numbers to 100 .. 4
- 1.5 Algebra: Different Ways to Show Numbers 5
- 1.6 Benchmark Numbers 6
- 1.7 Problem Solving Skill: Make Reasonable Estimates 7

Chapter 2: Number Patterns and Compare Numbers
- 2.1 Algebra: Use Symbols >, <, or = .. 8
- 2.2 Problem Solving Skill: Use a Model 9
- 2.3 Algebra: Hundred Chart and Skip-Counting Patterns 10
- 2.4 Even and Odd 11
- 2.5 Problem Solving Strategy: Look for a Pattern 12

Chapter 3: Tables and Graphs
- 3.1 Venn Diagrams 13
- 3.2 Take a Survey on a Tally Table ... 14
- 3.3 Use Data in Tables 15
- 3.4 Picture Graphs 16
- 3.5 Problem Solving Skill: Use Data from a Graph 17
- 3.6 Make a Bar Graph 18

Chapter 4: Addition Strategies
- 4.1 Algebra: Commutative and Identity Properties of Addition 19
- 4.2 Count On 20
- 4.3 Doubles and Doubles Plus One ... 21
- 4.4 Make a Ten 22
- 4.5 Algebra: Associative Property of Addition 23
- 4.6 Problem Solving Strategy: Make a Picture 24

Chapter 5: Subtraction Strategies
- 5.1 Count Back 25
- 5.2 Algebra: Fact Families 26
- 5.3 Algebra: Relate Addition to Subtraction 27
- 5.4 Algebra: Missing Addends 28
- 5.5 Equal and Not Equal 29
- 5.6 Algebra: Names for Numbers 30
- 5.7 Problem Solving Strategy: Make a Picture 31

Unit 2: 2-DIGIT ADDITION AND SUBTRACTION

Chapter 6: Explore 2-Digit Addition
- 6.1 Mental Math: Add Tens 32
- 6.2 Mental Math: Count on Tens and Ones 33
- 6.3 Regroup Ones and Tens 34
- 6.4 Model 2-Digit Addition 35
- 6.5 Problem Solving Strategy: Use Objects 36

▶ **Chapter 7: 2-Digit Addition**
   7.1 Add 1-Digit Numbers 37
   7.2 Add 2-Digit Numbers 38
   7.3 More 2-Digit Addition 39
   7.4 Rewrite 2-Digit Addition 40
   7.5 Estimate Sums 41
   7.6 Use Mental Math to Find Sums .. 42
   7.7 Problem Solving Skill: Too Much Information 43

▶ **Chapter 8: Explore 2-Digit Subtraction**
   8.1 Mental Math: Subtract Tens 44
   8.2 Mental Math: Count Back Tens and Ones 45
   8.3 Regroup Tens as Ones 46
   8.4 Model 2-Digit Subtraction 47
   8.5 Problem Solving Skill: Choose the Operation 48

▶ **Chapter 9: 2-Digit Subtraction**
   9.1 Subtract 1-Digit Numbers 49
   9.2 Subtract 2-Digit Numbers 50
   9.3 More 2-Digit Subtraction 51
   9.4 Rewrite 2-Digit Subtraction 52
   9.5 Estimate Differences 53
   9.6 Algebra: Use Addition to Check Subtraction 54
   9.7 Use Mental Math to Find Differences 55
   9.8 Problem Solving Skill: Choose the Computational Method 56

▶ **Chapter 10: Practice 2-Digit Addition and Subtraction**
   10.1 Different Ways to Add 57
   10.2 Practice 2-Digit Addition 58
   10.3 Column Addition 59
   10.4 Different Ways to Subtract 60
   10.5 Practice 2-Digit Subtraction 61
   10.6 Mixed Practice 62
   10.7 Problem Solving Skill: Too Much Information 63

▶ **Unit 3: MONEY AND TIME**

▶ **Chapter 11: Count and Use Money**
   11.1 Pennies, Nickels, Dimes, and Quarters 64
   11.2 Count Collections 65
   11.3 Make the Same Amounts 66
   11.4 Algebra: Same Amounts Using the Fewest Coins 67
   11.5 Make Change to $1.00 68
   11.6 Count Bills and Coins 69
   11.7 More Bills and Coins 70
   11.8 Problem Solving Strategy: Act it Out 71
   11.9 Make Change to $5.00 72
   11.10 Make Change to $10.00 73

▶ **Chapter 12: Tell and Understand Time**
   12.1 Time to the Hour 74
   12.2 Time to the Half-Hour 75
   12.3 Time to 15 Minutes 76
   12.4 Time to 5 Minutes 77
   12.5 Problem Solving Skill: Use a Model 78
   12.6 Hours, Days, Weeks, Months, Years 79
   12.7 Sequence Events 80

Unit 4: GEOMETRY, MEASUREMENT, AND FRACTIONS

Chapter 13: Plane Shapes
- 13.1 Plane Shapes 81
- 13.2 Algebra: Sort Plane Shapes 82
- 13.3 Plane Shapes with 4 Sides 83
- 13.4 More About Plane Shapes 84
- 13.5 Angles of Plane Shapes 85
- 13.6 Combine and Separate Shapes ... 86
- 13.7 Problem Solving Strategy: Use Objects 87

Chapter 14: Solid Figures
- 14.1 Solid Figures 88
- 14.2 Algebra: Sort Solid Figures 89
- 14.3 Compare Solid Figures & Plane Shapes 90
- 14.4 Combine Solid Figures 91
- 14.5 Take Apart Solid Figures 92
- 14.6 Problem Solving Skill: Use a Table 93

Chapter 15: Customary Measurement: Length and Temperature
- 15.1 Length 94
- 15.2 Measure to the Nearest Inch 95
- 15.3 Inch, Foot, and Yard 96
- 15.4 Inches and Feet 97
- 15.5 Feet and Yards 98
- 15.6 Fahrenheit Temperature 99
- 15.7 Problem Solving Skill: Make Reasonable Estimates 100
- 15.8 Problem Solving Skill: Choose the Measuring Tool and Unit ... 101

Chapter 16: Metric Measurement
- 16.1 Measure to the Nearest Centimeter 102
- 16.2 Explore Centimeters and Meters 103
- 16.3 Centimeter and Meter 104
- 16.4 Problem Solving Strategy: Guess and Check 105

Chapter 17: Fractions
- 17.1 Explore Fractions 106
- 17.2 Unit Fractions 107
- 17.3 Problem Solving Strategy: Use Objects 108
- 17.4 Other Fractions 109
- 17.5 Fractions Equal to 1 110
- 17.6 Unit Fractions of a Set 111
- 17.7 Other Fractions of a Set 112

Unit 5: EXPLORE GREATER NUMBERS AND OPERATIONS

Chapter 18: Explore Greater Numbers
- 18.1 Hundreds 113
- 18.2 Hundreds, Tens, and Ones 114
- 18.3 Place Value 115
- 18.4 Algebra: Different Ways to Show Numbers 116
- 18.5 Problem Solving Strategy: Use Objects 117
- 18.6 Explore Thousands 118
- 18.7 Explore 4-Digit Numbers 119
- 18.8 10, 100, 1,000 120

Chapter 19: Compare Greater Numbers
- 19.1 Algebra: Use Symbols >, <, and = ... 121
- 19.2 Missing Numbers to 1,000 122
- 19.3 Algebra: Skip-Count 123
- 19.4 Problem Solving Strategy: Look for a Pattern 124

Chapter 20: Add 3-Digit Numbers
- 20.1 Mental Math: Add Hundreds 125
- 20.2 Equal and Not Equal 126
- 20.3 Model 3-Digit Addition: Regroup Ones 127
- 20.4 Model 3-Digit Addition: Regroup Tens 128
- 20.5 More 3-Digit Addition 129
- 20.6 Add Money 130
- 20.7 Problem Solving Skill: Choose the Computational Method 131

Chapter 21: Subtract 3-Digit Numbers
- 21.1 Mental Math: Subtract Hundreds 132
- 21.2 Model 3-Digit Subtraction: Regroup Tens 133
- 21.3 Model 3-Digit Subtraction: Regroup Hundreds 134
- 21.4 More 3-Digit Subtraction 135
- 21.5 Subtract Money 136
- 21.6 Practice 3-Digit Addition and Subtraction 137
- 21.7 Problem Solving Skill: Solve Multistep Problems 138

Unit 6: MULTIPLICATION AND DIVISION

Chapter 22: Multiplication
- 22.1 Addition and Multiplication 139
- 22.2 Arrays 140
- 22.3 Algebra: Multiply in Any Order .. 141
- 22.4 Multiply by 1 and Multiply by 0 142
- 22.5 Skip-Count to Multiply 143
- 22.6 Problem Solving Skill: Make a Table 144
- 22.7 Use a Multiplication Table 145

Chapter 23: Division
- 23.1 Equal Groups: Size of Groups ... 146
- 23.2 Equal Groups: Number of Groups 147
- 23.3 Subtraction and Division 148
- 23.4 Problem Solving Strategy: Act it Out 149
- 23.5 Problem Solving Skill: Choose the Computational Method 150

Name _____

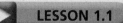 **LESSON 1.1**

Tens

Write how many tens.
Then write how many ones.

1.   __6__ tens = __60__ ones

2.   __4__ tens = __40__ ones

3.   __3__ tens = __30__ ones

4.   __2__ tens = __20__ ones

5.   __9__ tens = __90__ ones

▶ **Mixed Review**

Solve.

6. $8 + 2 = $ __10__ $3 + 2 = $ __5__ $6 + 2 = $ __8__

7. $4 + 4 = $ __8__ $6 + 1 = $ __7__ $8 + 1 = $ __9__

8. $5 + 2 = $ __7__ $3 + 3 = $ __6__ $9 + 1 = $ __10__

Name _____

LESSON 1.2

Tens and Ones

Write how many tens and ones in three different ways.

1.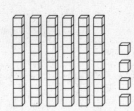

   __6__ tens __4__ ones = __64__
   __60__ + __4__ = __64__
             __64__

2.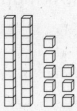

   __2__ tens __8__ ones = __28__
   __20__ + __8__ = __28__
             __28__

3.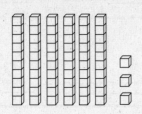

   __6__ tens __3__ ones = __63__
   __60__ + __3__ = __63__
             __63__

4.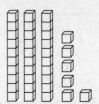

   __3__ tens __6__ ones = __36__
   __30__ + __6__ = __36__
             __36__

▶ **Mixed Review**

Write the sum.

5. 8 + 0 = __8__
6. 5 + 2 = __7__
7. 5 + 1 = __6__
8. 0 + 6 = __6__
9. 6 + 1 = __7__
10. 7 + 1 = __8__
11. 7 + 0 = __7__
12. 3 + 3 = __6__
13. 9 + 1 = __10__

Name _____

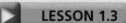 **LESSON 1.3**

Understand Place Value

Circle the value of the underlined digit.

1. 6<u>5</u> (5) or 50	2. <u>3</u>7 3 or (30)	3. <u>9</u>4 9 or (90)
4. 1<u>9</u> 1 or (10)	5. 4<u>3</u> (3) or 30	6. <u>5</u>1 5 or (50)
7. 8<u>7</u> (7) or 70	8. 1<u>2</u> (2) or 20	9. 7<u>5</u> (5) or 50
10. 3<u>9</u> (9) or 90	11. <u>8</u>7 8 or (80)	12. <u>9</u>1 9 or (90)

▶ **Mixed Review**

Solve.

13. 6 + 0 = __6__   6 + 4 = __10__   4 + 3 = __7__

14. 5 + 1 = __6__   5 + 4 = __9__   4 + 4 = __8__

15. 8 + 2 = __10__   3 + 3 = __6__   5 + 3 = __8__

Practice/Homework **PW3**

Name _____

LESSON 1.4

Read and Write Numbers to 100

Read the number.
Write the number in three different ways.

1. thirty-six

   _3_ tens _6_ ones
   _30_ + _6_
   _36_

2. fifty-five

   _5_ tens _5_ ones
   _50_ + _5_
   _55_

3. seventy-two

   _7_ tens _2_ ones
   _70_ + _2_
   _72_

4. eleven

   _1_ ten _1_ one
   _10_ + _1_
   _11_

5. twenty-two

   _2_ tens _2_ ones
   _20_ + _2_
   _22_

6. sixty-eight

   _6_ tens _8_ ones
   _60_ + _8_
   _68_

▶ **Mixed Review**

Solve.

7. 1 + 0 = _1_   8 + 2 = _10_   4 + 6 = _10_

8. 7 + 2 = _9_   2 + 4 = _6_   5 + 5 = _10_

9. 3 + 5 = _8_   1 + 9 = _10_   2 + 5 = _7_

PW4   Practice/Homework

Name _____

LESSON 1.5

Algebra: Different Ways to Show Numbers

Circle the correct ways to show each number.
Cross out and correct the other ways.

1. 34

   (30 + 4)

   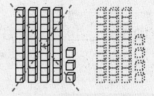

   ~~4 tens 3 ones~~
   3 tens 4 ones

2. 23

   (2 tens 3 ones)

   ~~20 + 5~~ 20 + 3

   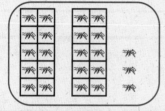

3. 20

   (20 + 0)

   ~~0 tens 2 ones~~
   2 tens 0 ones

4. 18

   ~~10 + 9~~ 10 + 8

   (1 ten 8 ones)

▶ **Mixed Review**

Write the sum.

5. 6 + 8 = __14__
8. 8 + 3 = __11__
11. 9 + 3 = __12__

6. 3 + 7 = __10__
9. 6 + 5 = __11__
12. 7 + 7 = __14__

7. 7 + 5 = __12__
10. 4 + 6 = __10__
13. 4 + 8 = __12__

Practice/Homework PW5

LESSON 1.6

Name _____

Benchmark Numbers

Find about how many there are in all.

1. This group has 10 leaves.

   About how many leaves are there in all? about __20__ leaves

2. This group has 10 shells.

   About how many shells are there in all? about __40__ shells

3. This group has 10 acorns.

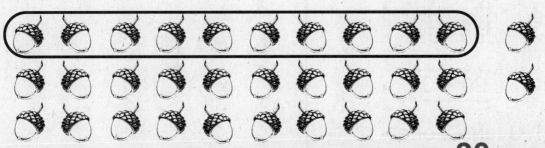

   About how many acorns are there in all? about __30__ acorns

▶ **Mixed Review**

Circle the value of the underlined digit.

4. 1̲8   5. 2̲4   6. 4̲9

   1 or (10)   (4) or 40   4 or (40)

PW6 Practice/Homework

Name _____

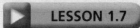

Problem Solving • Make Reasonable Estimates

Find the reasonable answer.
Circle the number that makes sense.

1. Lily has a few marbles. About how many marbles might she have?

   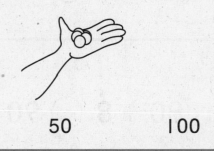

   (5)   50   100

2. Kim bought a small bag of apples. About how many apples might she have?

   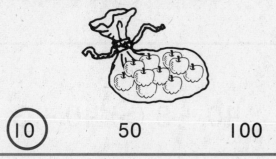

   (10)   50   100

3. Ann has a large collection of stickers. About how many stickers might she have?

   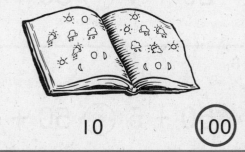

   5   10   (100)

4. Erica bought a box of pencils. About how many pencils might be in the box?

   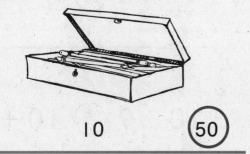

   5   10   (50)

5. Nick had some balloons. About how many balloons might he have?

   5   (10)   100

6. Jerry took out some books from the library. About how many books might that be?

   (5)   50   100

Practice/Homework   PW7

Name _____

LESSON 2.1

Algebra: Use Symbols: >, <, or =

Compare the numbers.
Write >, <, or =.

1. $70 + 4 \; \boxed{<} \; 80 + 9$

2. $90 + 8 \; \boxed{>} \; 80 + 7$

3. $40 + 8 \; \boxed{>} \; 40 + 3$

4. $80 + 8 \; \boxed{<} \; 90 + 9$

5. $8 \; \boxed{=} \; 8$

6. $20 + 4 \; \boxed{<} \; 30 + 8$

7. $10 + 9 \; \boxed{>} \; 10 + 6$

8. $50 + 5 \; \boxed{=} \; 50 + 5$

▶ **Mixed Review**

Solve.

9. $3 + 4 = \underline{7}$ $2 + 8 = \underline{10}$ $7 + 3 = \underline{10}$

10. $9 + 0 = \underline{9}$ $0 + 7 = \underline{7}$ $5 + 3 = \underline{8}$

11. $1 + 7 = \underline{8}$ $4 + 6 = \underline{10}$ $7 + 2 = \underline{9}$

Name _____

LESSON 2.2

Problem Solving • Use a Model

Use the number line to
solve the problem.
Find the nearest ten.

Think:
If a number is halfway
between two tens, use
the greater ten.

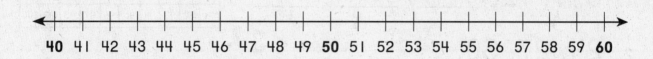

1. Gail reads 43 pages in her book.
   Does Gail read about 40 or
   50 pages?   about __40__ pages

2. The children make 58 paper birds.
   Do they make about 50 or 60
   paper birds?   about __60__ birds

3. Joan walks for 48 minutes.
   Does she walk about 40 or
   50 minutes?   about __50__ minutes

4. There are 55 apples in the box.
   Are there about 50 or 60 apples
   in the box?   about __60__ apples

5. There are 52 children in the school
   play. Are there about 50 or 60
   children in the play?   about __50__ children

6. There are 44 flowers in the garden.
   Are there about 40 or 50 flowers
   in the garden?   about __40__ flowers

Practice/Homework   PW9

Name _____

 LESSON 2.3

Algebra: Hundred Chart and Skip-Counting Patterns

Look for a pattern.
Write the missing numbers.
Use the hundred chart to help you.

1. 6, 9, 12, __15__, __18__, __21__

2. 24, 27, 30, __33__, __36__, __39__

3. 72, 75, 78, __81__, __84__, __87__

4. 55, 60, 65, __70__, __75__, __80__, __85__, __90__

5. 40, 45, 50, __55__, __60__, __65__, __70__, __75__

6. 30, 40, 50, __60__, __70__, __80__, __90__, __100__

7. 20, 30, 40, __50__, __60__, __70__, __80__, __90__

▶ **Mixed Review**

Tell how many tens and ones.

8. __7__ tens = __70__ ones

9. __4__ tens = __40__ ones

PW10 Practice/Homework

LESSON 2.4

Even and Odd

Draw to show the number as tens and ones.
Write **even** or **odd**.

1.   14

       even
_____

2.   23

       odd
_____

3.   37

       odd
_____

4.   18

       even
_____

▶ **Mixed Review**

Circle the correct ways to show the number.

5. 34   (3 tens 4 ones)   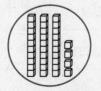   40 + 3

6. 21   (20 + 1)   10 + 2   (2 tens 1 one)

Practice/Homework   **PW11**

Name _____

LESSON 2.5

Problem Solving • Look For a Pattern

Find the pattern rule.
Complete the chart to solve.

1. How many wheels are on 6 wagons?

number of wagons	1	2	3	4	5	6
number of wheels	4	8	12	16	20	24

Count by __fours__.

There are __24__ wheels on 6 wagons.

2. How many vertices are on 7 triangles?

number of triangles	1	2	3	4	5	6	7
number of vertices	3	6	9	12	15	18	21

Count by __threes__.

There are __21__ vertices on 7 triangles.

3. How many fingers are on 8 hands?

number of hands	1	2	3	4	5	6	7	8
number of fingers	5	10	15	20	25	30	35	40

Count by __fives__.

There are __40__ fingers on 8 hands.

PW12 Practice/Homework

Name _____

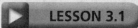

Venn Diagrams

Use the Venn diagram to answer the questions.

1. How many shapes are there in all?   **10** shapes

2. How many shapes are gray and are triangles?   **2** shapes

3. How many shapes are not gray?   **4** shapes

4. How many gray shapes are not triangles?   **4** shapes

▶ **Mixed Review**

Read the number. Then write the number.

5. sixty-eight **68**   6. forty-five **45**   7. seventeen **17**

8. eighty-two **82**   9. thirty-six **36**   10. fifty-one **51**

Name _____

LESSON 3.2

Take a Survey on a Tally Table

1. Take a survey. Ask ten classmates which pet they like the best. Fill in the tally table to show their answers.

Check children's work.

Pets We Like	
Pet	Tally
Cat	
Dog	
Fish	
Bird	
Hamster	

2. Which pet did the most children choose? _____

3. Did more children like dogs or birds the best? _____

4. How many children liked cats the best? _____

5. Which pet did the fewest children choose? _____

▶ **Mixed Review**

Fill in the missing numbers.

6. 21, 31, __41__, __51__, __61__, 71, __81__, __91__

7. 17, __27__, 37, __47__, __57__, __67__, 77, __87__

8. 86, 76, __66__, __56__, 46, __36__, __26__, __16__

PW14   Practice/Homework

Name _____   LESSON 3.3

Use Data in Tables

Use the tally tables to answer the questions.

Favorite Sandwich for Ken's Group	
Sandwich	Tally
peanut butter	IIII
chicken	II
tuna	I
ham and cheese	III

Favorite Sandwich for Ken's Class	
Sandwich	Tally
peanut butter	ЖHT III
chicken	III
tuna	IIII
ham and cheese	ЖHT II

1. Which sandwich did the most children in the group choose as their favorite? __peanut butter__

2. Which sandwich did the most children in the class choose as their favorite? __peanut butter__

3. Which sandwich did the fewest children in Ken's group choose? __tuna__

4. Could the survey of the group help you predict the results for the class? __yes__

 Mixed Review

Write **even** or **odd**.

5. 21 __odd__   6. 30 __even__   7. 12 __even__

8. 32 __even__   9. 6 __even__   10. 24 __even__

Practice/Homework PW15

Name _____

Picture Graphs

Use the picture graph to answer the questions.

How Children in Our Class Get to School											
Bus	🚌	🚌	🚌	🚌	🚌						
Car	🚗	🚗	🚗	🚗	🚗	🚗	🚗	🚗	🚗	🚗	🚗
Bike	🚲	🚲	🚲	🚲	🚲	🚲	🚲	🚲	🚲		

1. How many children ride the bus to school?   __5__ children

2. Which way do most of the children get to school?   __by car__

3. How many more children ride in a car to school than ride a bike?   __2__ more children

4. Write an addition or subtraction problem about the graph.   __Check children's work.__

 Mixed Review

Solve.

5. 8 + 3 = __11__   6. 7 + 4 = __11__   7. 5 + 6 = __11__
8. 5 + 8 = __13__   9. 4 + 9 = __13__   10. 6 + 7 = __13__

Name _____

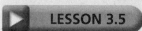

Problem Solving • Use Data from a Graph

Jamie's class made a bar graph to show the hobbies they like.

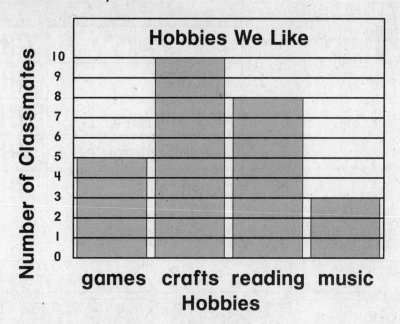

Use the bar graph to answer the questions.

1. Which hobby was chosen by the most children?   **crafts**

2. Which hobby was chosen by the fewest children?   music

3. How many children like games or reading?   13

4. How many children like crafts or music?   13

5. How many more children like reading than music?   5

6. How many more children like crafts than games?   5

Practice/Homework   PW17

Name _____

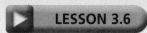

Make a Bar Graph

Alicia asked her classmates which of five evening activities they liked best. She made a tally table to keep track of their answers.

Evening Activities We Like	
Activity	Tally
read a book	𝄁𝄁𝄁𝄁𝄁 𝄁𝄁
watch TV	𝄁𝄁𝄁𝄁𝄁 𝄁𝄁𝄁
play with toys	𝄁𝄁𝄁𝄁
use the computer	𝄁𝄁𝄁𝄁𝄁 𝄁𝄁𝄁𝄁𝄁
play sports	𝄁𝄁𝄁𝄁𝄁 𝄁

1. Use Alicia's tally table to make a bar graph.

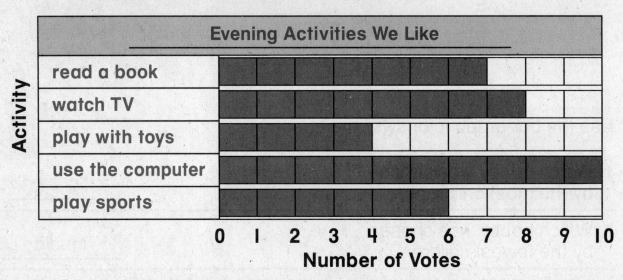

2. Use the bar graph to order the activities from the one that received the most votes to the one that received the fewest votes.

Most   use the computer
              watch TV
              read a book
              play sports
Fewest   play with toys

Name _____

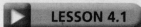 **LESSON 4.1**

Algebra: Commutative and Identity Properties of Addition

Write the sum.

1.   3   2
   + 2   + 3
   ───   ───
     5   5

2.   0   4
   + 4   + 0
   ───   ───
     4   4

3.   6   4
   + 4   + 6
   ───   ───
    10   10

4.   3   0
   + 0   + 3
   ───   ───
     3   3

5.   2   7
   + 7   + 2
   ───   ───
     9   9

6.   5   0
   + 0   + 5
   ───   ───
     5   5

7.   0   8
   + 8   + 0
   ───   ───
     8   8

8.   4   7
   + 7   + 4
   ───   ───
    11   11

▶ **Mixed Review**

Write >, <, or =.

9. $8 + 2 \; \boxed{>} \; 3 + 4$ 10. $5 + 2 \; \boxed{=} \; 2 + 5$ 11. $6 + 3 \; \boxed{=} \; 4 + 5$

Practice/Homework   **PW19**

Name _____

 LESSON 4.2

Count On

Circle the greater number.
Count on to find the sum.

1. ⑧ + 1 = __9__	2. ⑤ + 2 = __7__	3. 3 + ⑩ = __13__
4. 1 + ④ = __5__	5. ⑥ + 2 = __8__	6. ⑦ + 3 = __10__
7. 3 + ⑧ = __11__	8. ⑧ + 2 = __10__	9. 2 + ⑦ = __9__
10. ⑦ + 3 = __10__	11. 1 + ⑤ = __6__	12. ⑥ + 1 = __7__
13. ④ + 3 = __7__	14. 2 + ⑩ = __12__	15. 3 + ⑥ = __9__
16. ⑤ + 3 = __8__	17. ⑨ + 2 = __11__	18. 3 + ⑨ = __12__

▶ **Mixed Review**

Write >, <, or =.

19. 10 + 1 ⟨<⟩ 9 + 3   4 − 2 ⟨>⟩ 3 − 2

20. 10 + 1 ⟨=⟩ 7 + 4   7 + 6 ⟨>⟩ 9 + 3

PW20   Practice/Homework

Name _____

LESSON 4.3

Doubles and Doubles Plus One

Write the sums.

1. 4 + 4 = __8__, so 5 + 4 = __9__

2. 5 + 5 = __10__, so 5 + 6 = __11__

3. 2 + 2 = __4__, so 3 + 2 = __5__

4. 8 + 8 = __16__, so 8 + 9 = __17__

5. 1 + 1 = __2__, so 2 + 1 = __3__

6. 7 + 7 = __14__, so 7 + 8 = __15__

7. 3 + 3 = __6__, so 4 + 3 = __7__

8. 6 + 6 = __12__, so 6 + 7 = __13__

9. 9 + 9 = __18__, so 9 + 10 = __19__

▶ **Mixed Review**

Write the sums.

10. 5 + 2 = __7__   1 + 4 = __5__   3 + 9 = __12__

11. 7 + 1 = __8__   2 + 6 = __8__   7 + 3 = __10__

12. 3 + 4 = __7__   9 + 2 = __11__   8 + 1 = __9__

Practice/Homework **PW21**

Name _____

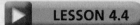 **LESSON 4.4**

Make a Ten

Use a ten frame and ⬤ to find the sum.

1. 7
  + 5

   12

   THINK: Start with 7. Borrow 3 from 5 to make a ten. 10 + 2 = 12

2. 8
  + 6

   14

3. 9
  + 1

   10

4. 3
  + 8

   11

5. 5
  + 7

   12

6. 7
  + 4

   11

7. 6
  + 8

   14

8. 9
  + 6

   15

9. 7
  + 6

   13

10. 7
  + 7

   14

11. 6
  + 9

   15

12. 2
  + 9

   11

13. 5
  + 8

   13

14. 8
  + 4

   12

15. 9
  + 2

   11

16. 7
  + 8

   15

17. 3
  + 7

   10

18. 8
  + 2

   10

19. 8
  + 8

   16

20. 8
  + 5

   13

21. 8
  + 3

   11

22. 9
  + 9

   18

23. 7
  + 9

   16

▶ **Mixed Review**

Circle the greater number. Count on to find the sum.

24. ⑧
  + 2

   10

25. ⑥
  + 1

    7

26. 3
  +⑥

    9

27. 1
  +⑦

    8

28. ②
  + 8

   10

PW22 Practice/Homework

Name _____

Algebra: Associative Property of Addition

Write the sum.

1. 2   2   2
    8 → 10 8 → 14 8 → 8
    +6 +6 +6 +2 +6 +8
       16   16   16

2. 3   3. 1   4. 5   5. 7   6. 2   7. 4
    1   6   8   5   6   3
    +3   +9   +2   +5   +4   +4
     7   16   15   17   12   11

8. 8   9. 5   10. 7   11. 4   12. 9   13. 2
    2   4   6   1   1   6
    +9   +4   +4   +4   +5   +2
    19   13   17   9   15   10

14. 3   15. 7   16. 9   17. 2   18. 8   19. 9
    4   3   0   4   3   5
    +1   +4   +9   +6   +2   +2
     8   14   18   12   13   16

▶ **Mixed Review**

Write the sum.

20. 7 + 1 = __8__   2 + 8 = __10__   9 + 3 = __12__

21. 4 + 4 = __8__   7 + 7 = __14__   3 + 3 = __6__

22. 5 + 6 = __11__   8 + 9 = __17__   9 + 10 = __19__

Practice/Homework PW23

LESSON 4.6

Name _____

Problem Solving • Make a Picture

Make a picture to solve. Write the number sentence.

Check children's work.

1. 9 brown bears and 7 black bears played. How many bears in all played?

   __9__ (+) __7__ (=) __16__ bears

2. 7 cats sat on the porch. Then 8 more cats joined them. Each cat has 4 legs. How many cats were on the porch?

   __7__ (+) __8__ (=) __15__ cats

3. 6 yellow fish and 8 orange fish swam in a fish tank. How many fish swam in the tank?

   __6__ (+) __8__ (=) __14__ fish

4. There were 7 children in the yard and 3 children in the house. There were 4 adults watching the children. How many children were there in all?

   __7__ (+) __3__ (=) __10__ children

PW24 Practice/Homework

Name _____

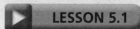 LESSON 5.1

Count Back

Count back to find the difference.

1. $8 - 1 = \underline{7}$ $4 - 2 = \underline{2}$ $6 - 1 = \underline{5}$

2. $5 - 2 = \underline{3}$ $9 - 3 = \underline{6}$ $10 - 2 = \underline{8}$

3.
```
   7      5      8      4      6
  -3     -1     -3     -1     -2
  ──     ──     ──     ──     ──
   4      4      5      3      4
```

4.
```
  10      9     11      7      3
  -3     -2     -2     -2     -2
  ──     ──     ──     ──     ──
   7      7      9      5      1
```

5.
```
   8      3      9     12      7
  -2     -1     -1     -1     -1
  ──     ──     ──     ──     ──
   6      2      8     11      6
```

6.
```
   3     10      6     11      5
  -2     -1     -3     -3     -3
  ──     ──     ──     ──     ──
   1      9      3      8      2
```

▶ **Mixed Review**

Write the missing numbers.

7. 25, 26, __27__, __28__, 29, __30__, __31__, 32

8. 12, __13__, __14__, __15__, 16, 17, 18, __19__, __20__

9. 63, 62, __61__, 60, __59__, __58__, 57, __56__, 55

Practice/Homework **PW25**

Name _____

LESSON 5.2

Algebra: Fact Families

Complete the fact families.   Order of facts may vary.

1. (14, 8, 6)

   8 + 6 = 14
   6 + 8 = 14
   14 − 6 = 8
   14 − 8 = 6

2. (7, 9, 16)

   9 + 7 = 16
   7 + 9 = 16
   16 − 7 = 9
   16 − 9 = 7

3. (7, 13, 6)

   6 + 7 = 13
   7 + 6 = 13
   13 − 7 = 6
   13 − 6 = 7

▶ **Mixed Review**

Write the sum.

4. 6 + 7 = __13__   5 + 6 = __11__   8 + 9 = __17__

5. 3 + 4 = __7__   7 + 8 = __15__   4 + 5 = __9__

PW26 Practice/Homework

Name _____

LESSON 5.3

Algebra: Relate Addition to Subtraction

Find the difference.
Write the addition fact to help you.   Order of addends may vary.

1. $\begin{array}{r} 10 \\ -\ 4 \\ \hline 6 \end{array}$ + $\boxed{4}$ = $\boxed{10}$, with $\boxed{6}$ above

2. $\begin{array}{r} 13 \\ -\ 5 \\ \hline 8 \end{array}$ + $\boxed{5}$ = $\boxed{13}$, with $\boxed{8}$ above

3. $\begin{array}{r} 9 \\ -\ 3 \\ \hline 6 \end{array}$ + $\boxed{3}$ = $\boxed{9}$, with $\boxed{6}$ above

4. $\begin{array}{r} 11 \\ -\ 7 \\ \hline 4 \end{array}$ + $\boxed{7}$ = $\boxed{11}$, with $\boxed{4}$ above

5. $\begin{array}{r} 14 \\ -\ 9 \\ \hline 5 \end{array}$ + $\boxed{9}$ = $\boxed{14}$, with $\boxed{5}$ above

6. $\begin{array}{r} 15 \\ -\ 8 \\ \hline 7 \end{array}$ + $\boxed{8}$ = $\boxed{15}$, with $\boxed{7}$ above

7. $\begin{array}{r} 7 \\ -\ 2 \\ \hline 5 \end{array}$ + $\boxed{2}$ = $\boxed{7}$, with $\boxed{5}$ above

8. $\begin{array}{r} 12 \\ -\ 3 \\ \hline 9 \end{array}$ + $\boxed{3}$ = $\boxed{12}$, with $\boxed{9}$ above

9. $\begin{array}{r} 11 \\ -\ 2 \\ \hline 9 \end{array}$ + $\boxed{2}$ = $\boxed{11}$, with $\boxed{9}$ above

▶ **Mixed Review**

Write <, >, or =.

10. $3 + 4\ \boxed{>}\ 1 + 3$ $10 + 10\ \boxed{<}\ 10 + 20$

11. $4 + 4\ \boxed{=}\ 5 + 3$ $20 + 20\ \boxed{<}\ 40 + 40$

Practice/Homework **PW27**

Name _____

▶ LESSON 5.4

Algebra: Missing Addends

Use addition and a related subtraction fact to find the missing addend.

1. $6 + \underline{8} = 14$ $14 - 6 = \underline{8}$

2. $\underline{7} + 5 = 12$ $12 - 5 = \underline{7}$

3. $9 + \underline{4} = 13$ $13 - 9 = \underline{4}$

4. $8 + \underline{8} = 16$ $16 - 8 = \underline{8}$

5. $\underline{7} + 6 = 13$ $13 - 6 = \underline{7}$

6. $\underline{9} + 9 = 18$ $18 - 9 = \underline{9}$

7. $8 + \underline{2} = 10$ $10 - 8 = \underline{2}$

▶ **Mixed Review**

Write the sum or difference.

8. $6 + 6 = \underline{12}$ $8 + 5 = \underline{13}$ $3 + 7 = \underline{10}$

9. $8 - 3 = \underline{5}$ $9 - 2 = \underline{7}$ $4 + 6 = \underline{10}$

0. $7 - 7 = \underline{0}$ $7 + 5 = \underline{12}$ $9 - 8 = \underline{1}$

LESSON 5.5

Name _____

Equal and Not Equal

Write **is equal to** or **is not equal to**. Then write = or ≠.

1. 20 + 6 __is equal to__ 26.
   20 + 6 (=) 26

2. 40 + 5 __is equal to__ 45.
   40 + 5 (=) 45

3. 60 + 7 __is not equal to__ 76.
   60 + 7 (≠) 76

4. 30 + 1 __is equal to__ 31.
   30 + 1 (=) 31

5. 40 + 2 __is not equal to__ 40.
   40 + 2 (≠) 40

6. 50 + 8 __is not equal to__ 60.
   50 + 8 (≠) 60

7. 10 + 9 __is equal to__ 19.
   10 + 9 (=) 19

8. 70 + 5 __is not equal to__ 85.
   70 + 5 (≠) 85

▶ **Mixed Review**

Write how many tens and ones.

9. 17
   __1__ ten __7__ ones

10. 36
    __3__ tens __6__ ones

11. 44
    __4__ tens __4__ ones

12. 21
    __2__ tens __1__ one

Name _____

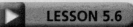

Algebra: Names for Numbers

Check children's work.

Write an addition and a subtraction name for each number.

1. [6]   3 + 3   9 − 3

2. [15] ___ + ___   ___ − ___

3. [4]   ___ + ___   ___ − ___

4. [16] ___ + ___   ___ − ___

5. [7]   ___ + ___   ___ − ___

6. [13] ___ + ___   ___ − ___

7. [18] ___ + ___   ___ − ___

8. [5]   ___ + ___   ___ − ___

▶ **Mixed Review**

Write even or odd.

9. 12 __even__   8 __even__   27 __odd__

10. 33 __odd__   40 __even__   15 __odd__

PW30 Practice/Homework

Name _____

LESSON 5.7

Problem Solving • Make a Picture.

Make a picture.
Write a number sentence to solve.

Check children's work.

1. Julie buys 13 apples. 5 of them are red and the rest are green. How many green apples does Julie buy.

   _13_ _5_ _8_
                     green apples

2. Mary has 6 dolls. Tasha has 4 dolls. How many more dolls does Mary have than Tasha?

   _6_ _4_ _2_
                 more dolls

3. Joel plants 7 carrot seeds. He also plants some flower seeds. He plants 13 seeds altogether. How many flower seeds does Joel plant?

   _13_ _7_ _6_
                flower seeds

4. Eddie has 16 oranges. He gives 8 of them away. How many oranges does Eddie have left?

   _16_ _8_ _8_
                oranges

Practice/Homework **PW31**

Name _____

LESSON 6.1

Mental Math: Add Tens

Add.

1. 1 ten + 2 tens = __3__ tens
   __10__ + __20__ = __30__

2. 3 tens + 2 tens = __5__ tens
   __30__ + __20__ = __50__

3. 3 tens + 3 tens = __6__ tens
   __30__ + __30__ = __60__

4. 0 tens + 2 tens = __2__ tens
   __0__ + __20__ = __20__

5. 4 tens + 3 tens = __7__ tens
   __40__ + __30__ = __70__

6. 1 ten + 3 tens = __4__ tens
   __10__ + __30__ = __40__

7. 4 tens + 4 tens = __8__ tens
   __40__ + __40__ = __80__

8. 6 ten + 3 tens = __9__ tens
   __60__ + __30__ = __90__

▶ **Mixed Review**

Write the difference.

9. 8 − 2 = __6__   12 − 1 = __11__   7 − 3 = __4__

10. 10 − 3 = __7__   5 − 2 = __3__   11 − 1 = __10__

PW32 Practice/Homework

Name _____

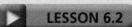 **LESSON 6.2**

Mental Math: Count On Tens and Ones

Count on to add.

1.
30	75	61	54	18
+39	+ 3	+30	+ 2	+20
69	78	91	56	38

2.
1	44	67	83	2
+29	+20	+10	+ 3	+41
30	64	77	86	43

3.
90	3	54	74	38
+ 3	+18	+30	+ 2	+10
93	21	84	76	48

4.
2	21	36	55	67
+59	+ 3	+20	+10	+ 2
61	24	56	65	69

▶ **Mixed Review**

What comes next? Write the number.

5. 3, 6, 9, __12__   7, 8, 9, __10__   22, 24, 26, __28__

6. 25, 30, 35, __40__   20, 30, 40, __50__   10, 12, 14, __16__

Practice/Homework   PW33

LESSON 6.3

Name _____

Regroup Ones and Tens

Use Workmat 3 and ▭▭▭▭ ▫ .

	Show.	Add.	Do you need to regroup? Circle Yes or No.	How many tens and ones?
1.	16	7	(Yes) No	_2_ tens _3_ ones
2.	34	7	(Yes) No	_4_ tens _1_ ones
3.	46	4	(Yes) No	_5_ tens _0_ ones
4.	63	5	Yes (No)	_6_ tens _8_ ones
5.	38	5	(Yes) No	_4_ tens _3_ ones

▶ **Mixed Review**

Write the difference.

6. $13 - 7 = \underline{6}$ 7. $10 - 10 = \underline{0}$ 8. $14 - 7 = \underline{7}$

9. $15 - 8 = \underline{7}$ 10. $16 - 8 = \underline{8}$ 11. $12 - 5 = \underline{7}$

Practice/Homework

Name _____

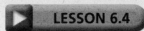

Model 2-Digit Addition

Use Workmat 3 and ▭▭▭▭▭▭ ▫.
Draw the regrouping if you need to. Then add.

Check children's work.

1.
tens	ones
☐	
4	2
+ 1	5
5	7

2.
tens	ones
1	
2	3
+ 1	7
4	0

3.
tens	ones
☐	
3	8
+ 1	1
4	9

4.
tens	ones
1	
2	6
+ 2	4
5	0

 Mixed Review

Write **even** or **odd**.

5. 17 __odd__   6. 18 __even__   7. 19 __odd__

8. 20 __even__   9. 21 __odd__   10. 22 __even__

Practice/Homework **PW35**

Name _____

LESSON 6.5

Problem Solving • Use Objects

Use Workmat 3 and ▭▭▭▭▭▭▭ ▫.
Add. Regroup if you need to.
Write the sum.

1. The sports store sold 13 mitts last week and 17 mitts this week. How many mitts were sold?

   __30__ mitts

tens	ones
1	
1	3
+1	7
3	0

2. There are 20 baseball bats for sale on the shelf. There are 19 bats in the back room. How many bats are for sale in all?

   __39__ bats

tens	ones
2	0
+1	9
3	9

3. One box holds 18 baseballs. Another box holds 23 baseballs. How many baseballs are there in all?

   __41__ baseballs

tens	ones
1	
1	8
+2	3
4	1

4. 19 children bought baseball caps on Monday. 16 children bought caps on Tuesday. How many caps were sold in all?

   __35__ caps

tens	ones
1	
1	9
+1	6
3	5

PW36 Practice/Homework

Add 1-Digit Numbers

Use Workmat 3 and ▭ . **Add.**

THINK: Do I need to regroup 10 ones as 1 ten?

#	Tens	Ones
1.	[1] 3	8 +7
	4	**5**

#	Tens	Ones
2.	[1] 1	5 +5
	2	**0**

#	Tens	Ones
3.	[ ] 5	2 +4
	5	**6**

#	Tens	Ones
4.	[ ] 2	5 +4
	2	**9**

#	Tens	Ones
5.	[1] 4	4 +6
	5	**0**

#	Tens	Ones
6.	[1] 3	9 +2
	4	**1**

#	Tens	Ones
7.	[ ] 5	1 +7
	5	**8**

#	Tens	Ones
8.	[1] 8	7 +4
	9	**1**

#	Tens	Ones
9.	[1] 3	8 +4
	4	**2**

#	Tens	Ones
10.	[1] 5	1 +9
	6	**0**

#	Tens	Ones
11.	[ ] 8	3 +5
	8	**8**

#	Tens	Ones
12.	[1] 6	7 +7
	7	**4**

▶ **Mixed Review**

Count back to find the difference.

13. 11 − 3 = **8** 14. 10 − 1 = **9** 15. 4 − 2 = **2**

16. 6 − 1 = **5** 17. 12 − 3 = **9** 18. 7 − 1 = **6**

Name _____

LESSON 7.2

Add 2-Digit Numbers

Use Workmat 3 and ⬚⬚⬚⬚⬚ ▫ .
Add. Regroup if you need to.

1.	Tens	Ones
	[1]	
	5	9
+	1	4
	7	3

2.	Tens	Ones
	[1]	
	3	5
+	1	9
	5	4

3.	Tens	Ones
	[1]	
	5	6
+	1	5
	7	1

4.	Tens	Ones
	[1]	
	4	4
+		8
	5	2

5.	Tens	Ones
	[1]	
	2	5
+	2	8
	5	3

6.	Tens	Ones
	[ ]	
	2	3
+	3	4
	5	7

7.	Tens	Ones
	[1]	
	4	9
+	4	1
	9	0

8.	Tens	Ones
	[1]	
	1	4
+	2	9
	4	3

9.	Tens	Ones
	[1]	
	2	7
+	3	7
	6	4

10.	Tens	Ones
	[1]	
	4	8
+	2	9
	7	7

11.	Tens	Ones
	[ ]	
	2	6
+	2	3
	4	9

12.	Tens	Ones
	[1]	
	3	6
+		8
	4	4

▶ **Mixed Review**

Write the number.

13. 5 tens 5 ones = __55__

14. 3 tens 7 ones = __37__

15. 6 tens 3 ones = __63__

16. 4 tens 5 ones = __45__

Name _____

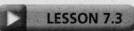

More 2-Digit Addition

Add.

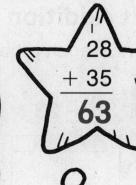
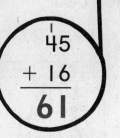

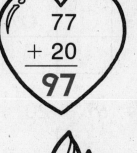

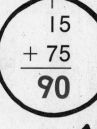

▶ **Mixed Review**

Write the missing numbers.

1. 82, __72__, __62__, 52, 42, __32__, 22, __12__

2. 97, 87, __77__, 67, __57__, __47__, __37__, 27

Practice/Homework   PW39

Name _____

LESSON 7.4

Rewrite 2-Digit Addition

Rewrite the numbers in each problem.
Then add.

1. 34 + 27

Tens	Ones
1	
3	4
+ 2	7
6	1

2. 36 + 23

Tens	Ones
3	6
+ 2	3
5	9

3. 27 + 16

Tens	Ones
1	
2	7
+ 1	6
4	3

4. 58 + 24

Tens	Ones
1	
5	8
+ 2	4
8	2

5. 23 + 52

```
  2 | 3
+ 5 | 2
-------
  7 | 5
```

6. 64 + 9

```
  1 |
  6 | 4
+   | 9
-------
  7 | 3
```

7. 67 + 20

```
  6 | 7
+ 2 | 0
-------
  8 | 7
```

8. 28 + 17

```
  1 |
  2 | 8
+ 1 | 7
-------
  4 | 5
```

9. 45 + 39

```
  1 |
  4 | 5
+ 3 | 9
-------
  8 | 4
```

10. 6 + 37

```
  1 |
    | 6
+ 3 | 7
-------
  4 | 3
```

11. 29 + 19

```
  1 |
  2 | 9
+ 1 | 9
-------
  4 | 8
```

12. 53 + 18

```
  1 |
  5 | 3
+ 1 | 8
-------
  7 | 1
```

▶ **Mixed Review**

Subtract.

13. 12 − 7 = __5__
14. 10 − 3 = __7__
15. 8 − 4 = __4__
16. 15 − 5 = __10__
17. 11 − 4 = __7__
18. 9 − 9 = __0__
19. 12 − 1 = __11__
20. 13 − 6 = __7__
21. 14 − 5 = __9__

Estimate Sums

LESSON 7.5

Add. Then use the number line to find the nearest ten. Estimate to see if your answer makes sense.

THINK: If a number is halfway between two tens, use the greater ten.

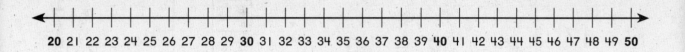

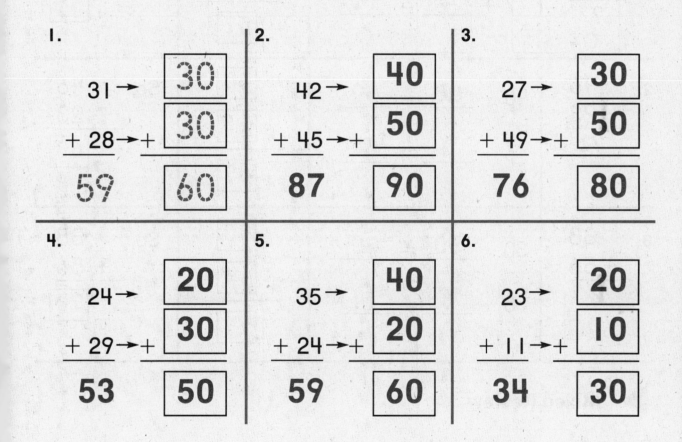

1. 31 → 30; + 28 → + 30; 59, 60
2. 42 → 40; + 45 → + 50; 87, 90
3. 27 → 30; + 49 → + 50; 76, 80
4. 24 → 20; + 29 → + 30; 53, 50
5. 35 → 40; + 24 → + 20; 59, 60
6. 23 → 20; + 11 → + 10; 34, 30

▶ **Mixed Review**

Circle the names for each number.

7. |11| (4 + 7) (12 − 1) 5 + 5 (8 + 3) 15 − 2
8. |12| 14 − 1 (4 + 8) (9 + 3) (15 − 3) (12 + 0)

Use Mental Math to Find Sums

Use mental math to add.

	Add the tens. Add the ones.	Add.	
1. 17 + 34 ?	10 + __30__ = __40__ 7 + __4__ = __11__	40 + 11 ――― 51	So, 17 + 34 ――― 51
2. 46 + 23 ?	40 + __20__ = __60__ 6 + __3__ = __9__	60 + 9 ――― 69	So, 46 + 23 ――― 69
3. 38 + 56 ?	30 + __50__ = __80__ 8 + __6__ = __14__	80 + 14 ――― 94	So, 38 + 56 ――― 94

▶ **Mixed Review**

Find the sum.

4. 3 + 9 = __12__　　5. 6 + 4 = __10__　　6. 7 + 7 = __14__

7. 7 + 8 = __15__　　8. 6 + 6 = __12__　　9. 8 + 8 = __16__

Name _____

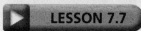

Problem Solving • Too Much Information

Draw a line through any information you do not need. Then solve.

THINK: Is there any information I do not need?

1. Hal swam 42 laps on Monday and 39 laps on Tuesday. ~~He swam the butterfly stroke for 7 laps.~~ How many laps did Hal swim on Monday and Tuesday?

   __81__ laps

   ```
     1
    42
   +39
   ---
    81
   ```

2. Last week Mrs. Jones baked 25 loaves of rye bread and 38 loaves of wheat bread. ~~She also baked 17 muffins.~~ How many loaves of bread did Mrs. Jones bake?

   __63__ loaves

   ```
     1
    25
   +38
   ---
    63
   ```

3. There are 17 cars parked on one side of the street. The are 21 cars parked on the other side. How many cars are parked on the street?

   __38__ cars

   ```
    17
   +21
   ---
    38
   ```

4. Tyrone put 17 pennies into his right pocket. He put 23 pennies into his left pocket. ~~He still had 4 pennies on his desk.~~ How many pennies did Tyrone put into his pockets?

   __40__ pennies

   ```
     1
    17
   +23
   ---
    40
   ```

Practice/Homework PW43

Name _____

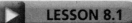

Mental Math: Subtract Tens

Subtract. Write the missing numbers.

1. 3 tens
    − 2 tens
    ──────
    1 ten

    30
    − 20
    ──
    10

2. 8 tens
    − 4 tens
    ──────
    4 tens

    80
    − 40
    ──
    40

3. 9 tens
    − 7 tens
    ──────
    2 tens

    90
    − 70
    ──
    20

4. 6 tens
    − 1 ten
    ──────
    5 tens

    60
    − 10
    ──
    50

5. 5 tens
    − 0 tens
    ──────
    5 tens

    50
    − 0
    ──
    50

6. 7 tens
    − 2 tens
    ──────
    5 tens

    70
    − 20
    ──
    50

▶ **Mixed Review**

Add.

7.
    35   24   19   48
    + 54   + 17   + 19   + 24
    ────   ────   ────   ────
    89   41   38   72

8.
    62   28   54   47
    + 29   + 12   + 27   + 32
    ────   ────   ────   ────
    91   40   81   79

PW44   Practice/Homework

Name _____

LESSON 8.2

Mental Math: Count Back Tens and Ones

Count back to subtract.

1.	66 − 20 **46**	52 − 40 **12**	77 − 3 **74**	22 − 10 **12**
2.	48 − 3 **45**	65 − 30 **35**	89 − 70 **19**	99 − 2 **97**
3.	36 − 3 **33**	44 − 20 **24**	18 − 3 **15**	59 − 10 **49**
4.	35 − 2 **33**	78 − 30 **48**	42 − 10 **32**	87 − 1 **86**

▶ **Mixed Review**

Solve.

5. 15 − 10 = **5**   12 − 8 = **4**   16 − 13 = **3**

6. 11 − 7 = **4**   12 − 6 = **6**   15 − 8 = **7**

7. 17 − 5 = **12**   15 − 13 = **2**   14 − 11 = **3**

8. 16 − 9 = **7**   14 − 7 = **7**   13 − 10 = **3**

Practice/Homework   PW45

Name _____

LESSON 8.3

Regroup Tens as Ones

Use Workmat 3 and ▭▭▭▭▭▭▭▭▭▭ ▫.

	Show.	Subtract.	Do you need to regroup? Circle Yes or No.	How many tens and ones are left?
1.	24	8	(Yes) No	__1__ tens __6__ ones
2.	32	5	(Yes) No	__2__ tens __7__ ones
3.	23	9	(Yes) No	__1__ tens __4__ ones
4.	70	8	(Yes) No	__6__ tens __2__ ones
5.	55	2	Yes (No)	__5__ tens __3__ ones

▶ **Mixed Review**

Add.

6. $7 + 6 =$ __13__   7. $9 + 2 =$ __11__   8. $8 + 6 =$ __14__

9. $8 + 8 =$ __16__   10. $4 + 8 =$ __12__   11. $9 + 5 =$ __14__

12. $5 + 7 =$ __12__   13. $11 + 5 =$ __16__   14. $3 + 7 =$ __10__

PW46   Practice/Homework

Name _____

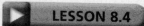 LESSON 8.4

Model 2-Digit Subtraction

Use Workmat 3 and ▭▭ ▫. Draw the regrouping.
Then subtract.

1. | tens | ones |
    |---|---|
    | 2 | 12 |
    | 3̸ | 2̸ |
    | 2 | 7 |
    | | 5 |

2. | tens | ones |
    |---|---|
    | 1 | 16 |
    | 2̸ | 6̸ |
    | 1 | 8 |
    | | 8 |

3. | tens | ones |
    |---|---|
    | ☐ | ☐ |
    | 4 | 6 |
    | 3 | 6 |
    | 1 | 0 |

4. | tens | ones |
    |---|---|
    | 1 | 12 |
    | 2̸ | 2̸ |
    | 1 | 3 |
    | | 9 |

▶ **Mixed Review**

Subtract.

5. 12 − 7 = __5__   16 − 7 = __9__   15 − 7 = __8__

6. 13 − 5 = __8__   17 − 9 = __8__   14 − 7 = __7__

Problem Solving • Choose the Operation

Use Workmat 3 and ▭▭▭▭▭▭▭▭▭▭ ▭.
Add or subtract. Write the sum or difference.

1. Josh put 12 toy cars on a shelf and 12 toy cars in a box. How many toy cars does he have in all?

   __24__ toy cars

tens	ones
1	2
1	2
2	4

   (+)

2. Cara is skating with 24 girls. Then her mom drives 5 of the girls home. How many girls are left to skate with Cara?

   __19__ girls

tens	ones
1	14
2̶	4̶
	5
1	9

   (−)

3. There are 16 toy train cars on the track. Then 4 of them fall off. How many train cars are left on the track?

   __12__ train cars

tens	ones
1	6
	4
1	2

   (−)

4. Jack got 16 rubber dinosaurs for his birthday. He already had 15 dinosaurs. How many rubber dinosaurs does he have now?

   __31__ dinosaurs

tens	ones
1	
1	6
1	5
3	1

   (+)

Name _____

Subtract 1-Digit Numbers

Use Workmat 3 and ▭▭▭▭▭▭▭▭▭▭ ▯ . Subtract.

LESSON 9.1

1. tens: 3̶[4] 7̶[17] − 8 = **39** (ones)
2. tens: 2 ones: 9 − 5 = **24**
3. tens: 1̶[2] ones: 5̶[15] − 7 = **18**
4. tens: 2̶[3] ones: 3̶[13] − 6 = **27**

5. tens: 7̶[8] ones: 2̶[12] − 5 = **77**
6. tens: 4 ones: 7 − 7 = **40**
7. tens: 1 ones: 4 − 2 = **12**
8. tens: 5̶[6] ones: 1̶[11] − 5 = **56**

9. tens: 3̶[4] ones: 2̶[12] − 8 = **34**
10. tens: 6 ones: 8 − 5 = **63**
11. tens: 2̶[3] ones: 5̶[15] − 9 = **26**
12. tens: 8̶[9] ones: 1̶[11] − 7 = **84**

▶ **Mixed Review**

Add.

13. 17 + 9 = **26**   10 + 37 = **47**   22 + 28 = **50**

14. 66 + 15 = **81**   45 + 15 = **60**   52 + 23 = **75**

Practice/Homework PW49

Subtract 2-Digit Numbers

Use Workmat 3 and ▭▭▭▭ ▫. Subtract.

THINK: When there are not enough ones, regroup 1 ten as 10 ones.

1. | tens | ones |
   |---|---|
   | 3̶ 4̶ | 1̶8̶ 8̶ |
   | − 1 | 9 |
   | 2 | 9 |

2. | tens | ones |
   |---|---|
   | ☐ 8 | ☐ 8 |
   | − 1 | 6 |
   | 7 | 2 |

3. | tens | ones |
   |---|---|
   | 5̶ 6̶ | 1̶1̶ 1̶ |
   | − 1 | 5 |
   | 4 | 6 |

4. | tens | ones |
   |---|---|
   | 1̶ 2̶ | 1̶3̶ 3̶ |
   | − | 8 |
   | 1 | 5 |

5. | tens | ones |
   |---|---|
   | 8̶ 9̶ | 1̶0̶ 0̶ |
   | − 2 | 5 |
   | 6 | 5 |

6. | tens | ones |
   |---|---|
   | ☐ 3 | ☐ 5 |
   | − 1 | 5 |
   | 2 | 0 |

7. | tens | ones |
   |---|---|
   | 6̶ 7̶ | 1̶6̶ 6̶ |
   | − | 7 |
   | 3 | 9 |

8. | tens | ones |
   |---|---|
   | ☐ 5 | ☐ 9 |
   | − 4 | 5 |
   | 1 | 4 |

9. | tens | ones |
   |---|---|
   | 6̶ 7̶ | 1̶3̶ 3̶ |
   | − 3 | 8 |
   | 3 | 5 |

10. | tens | ones |
    |---|---|
    | 2̶ 3̶ | 1̶4̶ 4̶ |
    | − 1 | 8 |
    | 1 | 6 |

11. | tens | ones |
    |---|---|
    | 6̶ 7̶ | 1̶7̶ 7̶ |
    | − 2 | 9 |
    | 4 | 8 |

12. | tens | ones |
    |---|---|
    | 5̶ 6̶ | 1̶4̶ 4̶ |
    | − 3 | 5 |
    | 2 | 9 |

▶ **Mixed Review**

Count back to find the difference.

13. 94 − 10 = __84__   77 − 10 = __67__   73 − 20 = __53__

14. 49 − 30 = __19__   62 − 20 = __42__   40 − 10 = __30__

More 2-Digit Subtraction

Subtract. Regroup if you need to.

1. tens: ³⁄4, ones: ¹⁵⁄5; − 2, 8; = 1, 7
2. tens: 5, ones: 9; − 1, 9; = 4, 0
3. tens: ⁵⁄6, ones: ¹⁵⁄5; − 4, 7; = 1, 8
4. tens: ⁸⁄9, ones: ¹⁴⁄4; − 2, 8; = 6, 6

5. tens: ⁶⁄7, ones: ¹⁷⁄7; − 1, 9; = 5, 8
6. tens: 5, ones: 3; − 2, 0; = 3, 3
7. tens: ³⁄4, ones: ¹⁰⁄0; − 1, 3; = 2, 7
8. tens: 7, ones: ¹³⁄3; − 2, 8; = 5, 5

9. tens: ⁴⁄5, ones: ¹¹⁄1; − 3, 2; = 1, 9
10. tens: ⁸⁄9, ones: ¹⁵⁄5; − 3, 7; = 5, 8
11. tens: 7, ones: ¹⁷⁄7; − 1, 8; = 6, 9
12. tens: ⁴⁄5, ones: ¹⁰⁄0; − 2, 5; = 2, 5

▶ **Mixed Review**

Write the missing numbers.

13. 14, __16__, 18, 20, __22__, __24__, 26, __28__

14. 9, __12__, __15__, 18, __21__, 24, __27__, 30

Name _____

LESSON 9.4

Rewrite 2-Digit Subtraction

Rewrite the numbers. Then subtract.

1. 61 − 37	2. 77 − 71	3. 95 − 48	4. 40 − 29
5 | 11 6̶ | 1̶ − 3 | 7 2 | 4	7 | 7 − 7 | 1 6	8 | 15 9̶ | 5̶ − 4 | 8 4 | 7	3 | 10 4̶ | 0̶ − 2 | 9 1 | 1
5. 64 − 27	6. 62 − 22	7. 33 − 15	8. 62 − 33
5 | 14 6̶ | 4̶ − 2 | 7 3 | 7	6 | 2 − 2 | 2 4 | 0	2 | 13 3̶ | 3̶ − 1 | 5 1 | 8	5 | 12 6̶ | 2̶ − 3 | 3 2 | 9
9. 63 − 37	10. 86 − 8	11. 71 − 69	12. 82 − 34
5 | 13 6̶ | 3̶ − 3 | 7 2 | 6	7 | 16 8̶ | 6̶ − | 8 7 | 8	6 | 11 7̶ | 1̶ − 6 | 9 | 2	7 | 12 8̶ | 2̶ − 3 | 4 4 | 8

▶ **Mixed Review**

Write the missing numbers.

13. 30, 40, __50__, 60, __70__, 80, 90

14. 14, __24__, 34, __44__, 54, __64__, 74

PW52 Practice/Homework

Name _____

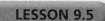

Estimate Differences

Subtract. Then use the number line to find the nearest ten. Estimate the difference to see if your answer makes sense.

THINK: If a number is halfway between two tens, use the greater ten.

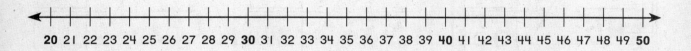

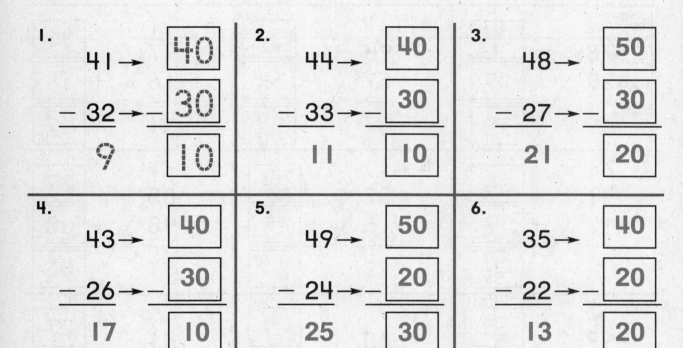

▶ **Mixed Review**

Write **even** or **odd**.

7. 10 __even__
8. 8 __even__
9. 15 __odd__

10. 17 __odd__
11. 6 __even__
12. 13 __odd__

13. 27 __odd__
14. 34 __even__
15. 67 __odd__

Practice/Homework PW53

Name _____

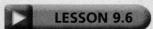

Algebra: Use Addition to Check Subtraction

Subtract. Add to check.

1. 56
   −11
   ───
   45   + 45 11 ── 56

2. ²¹⁴
   3̶4̶
   −16
   ───
   18   + 18 16 ── 34

3. 19
   −11
   ───
   8   + 8 11 ── 19

4. ⁶¹⁸
   7̶8̶
   −29
   ───
   49   + 49 29 ── 78

5. ⁸¹⁴
   9̶4̶
   −57
   ───
   37   + 37 57 ── 94

6. 47
   −16
   ───
   31   + 31 16 ── 47

7. ³¹¹
   4̶1̶
   −17
   ───
   24   + 24 17 ── 41

8. 37
   −15
   ───
   22   + 22 15 ── 37

9. ⁷¹⁵
   8̶5̶
   −48
   ───
   37   + 37 48 ── 85

10. 99
    −27
    ───
    72   + 72 27 ── 99

11. ⁷¹⁵
    8̶5̶
    −76
    ───
    9   + 9 76 ── 85

12. ⁴¹¹
    5̶1̶
    −24
    ───
    27   + 27 24 ── 51

▶ **Mixed Review**

Add.

13. 24
    + 3
    ───
    27

14. 33
    + 8
    ───
    41

15. 21
    +37
    ───
    58

16. 38
    +17
    ───
    45

PW54 Practice/Homework

Use Mental Math to Find Differences

Use mental math to subtract.

	Add the same number to both numbers.	Subtract.	
1. 45 −29 ?	45 + __1__ = __46__ 29 + __1__ = __30__	46 − 30 ――― 16	So, 45 −29 ――― 16
2. 63 −38 ?	63 + __2__ = __65__ 38 + __2__ = __40__	65 − 40 ――― 25	So, 63 −38 ――― 25

Try subtracting these numbers in your head.

3. 43
   −27
   ―――
   16

4. 65
   −46
   ―――
   19

5. 86
   −68
   ―――
   18

6. 71
   −44
   ―――
   27

7. 52
   −35
   ―――
   17

8. 48
   −19
   ―――
   29

9. 47
   −28
   ―――
   19

10. 73
   −35
   ―――
   38

▶ **Mixed Review**

Count on to add.

11. 57
   + 2
   ―――
   59

12. 40
   +37
   ―――
   77

13. 38
   +20
   ―――
   58

14. 62
   + 3
   ―――
   65

Name _____

LESSON 9.8

Problem Solving • Choose the Computational Method

Choose a method and solve the problem.

Use paper and pencil.	Use a calculator.	Count back.	Use base ten blocks.

1. Anna collects sports stamps. She has 67 stamps. She gives 14 away. How many stamps does Anna have left? __53__ stamps	Check children's work.
2. A page in a stamp book holds 32 stamps. Mark has 20 stamps. How many more stamps does he need to fill the page? __12__ stamps	Check children's work.
3. Alma received 21 stamps on her birthday. Her grandmother gave her 12 of them. How many stamps were not from her grandmother? __9__ stamps	Check children's work.
4. Alfredo has 83 stamps in his collection. He sends 29 stamps to his cousin. How many stamps does Alfredo have left? __54__ stamps	Check children's work.

PW56 Practice/Homework

Name _____

LESSON 10.1

Different Ways to Add

Choose the best way to add. Find the sum.

1. 57 + 30 = 87
2. 14 + 13 = 27
3. 62 + 4 = 66
4. 44 + 36 = 80
5. 29 + 8 = 37

6. 21 + 40 = 61
7. 18 + 18 = 36
8. 24 + 5 = 29
9. 65 + 10 = 75
10. 12 + 48 = 60

11. 81 + 10 = 91
12. 26 + 50 = 76
13. 38 + 13 = 51
14. 52 + 11 = 63
15. 37 + 36 = 73

16. 14 + 67 = 81
17. 6 + 40 = 46
18. 18 + 3 = 21
19. 48 + 27 = 75
20. 49 + 30 = 79

▶ **Mixed Review**

Count back to subtract.

21. 48 − 10 = __38__
22. 70 − 10 = __60__
23. 91 − 10 = __81__

24. 37 − 10 = __27__
25. 89 − 10 = __79__
26. 16 − 10 = __6__

Practice/Homework

Name _____

LESSON 10.2

Practice 2-Digit Addition

Toss a number cube with the numbers 1, 2, and 3 to fill in the boxes. Then add. **Check children's work.**

1. ☐ 6
   +☐ 4
   ─────

2. 1 ☐
   + 4 ☐
   ─────

3. ☐ 2
   + 6 ☐
   ─────

4. ☐ 8
   + 1 ☐
   ─────

5. 2 ☐
   + 5 ☐
   ─────

6. ☐ 7
   + 2 ☐
   ─────

7. 3 ☐
   +   9
   ─────

8. 2 ☐
   +☐ 7
   ─────

9. 3 ☐
   + 5 ☐
   ─────

10. ☐ 9
   + 1 ☐
   ─────

11. ☐ 5
   +☐ 5
   ─────

12. 5 ☐
   +   8
   ─────

▶ **Mixed Review**

Write how many tens and ones.

13. 15   __1__ ten __5__ ones

14. 49   __4__ tens __9__ ones

15. 31   __3__ tens __1__ one

16. 28   __2__ tens __8__ ones

Name _____

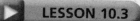 LESSON 10.3

Column Addition

Add.

1. 34
    16
    +25
    ----
    75   (with 10 and 15 grouping marks)

2. 18
    31
    +41

    90

3.   7
    57
    +11

    75

4. 26
    26
    + 2

    54

5. 15
    19
    +33

    67

6. 41
     9
    +17

    67

7. 55
    15
    + 8

    78

8. 29
    12
    +24

    65

9. 33
    16
    +49

    98

10. 18
    30
    + 5

    53

11. 21
    56
    +12

    89

12. 6
    16
    +26

    48

13. 72
    16
    +11

    99

14. 16
    29
    +31

    76

15. 41
     4
    +35

    80

16. 17
     3
    +28

    48

▶ **Mixed Review**

Write the missing addend.

17. 8 + __7__ = 15   15 − 8 = __7__

18. __6__ + 6 = 12   12 − 6 = __6__

Practice/Homework   **PW59**

Different Ways to Subtract

Choose the best way to subtract. Find the difference.
Then use the code to read the message.

1 – A	2 – B	3 – C	4 – D	5 – E
6 – F	7 – G	8 – H	9 – I	10 – J
11 – K	12 – L	13 – M	14 – N	15 – O
16 – P	17 – Q	18 – R	19 – S	20 – T
21 – U	22 – V	23 – W	24 – X	25 – Y
26 – Z				

What's gray and has a trunk?

1. 30 − 29 = **1**   **A**
2. 26 − 13 = **13**   **M**
3. 18 − 3 = **15**   **O**
4. 50 − 29 = **21**   **U**
5. 44 − 25 = **19**   **S**
6. 68 − 63 = **5**   **E**

7. 19 − 4 = **15**   **O**
8. 42 − 28 = **14**   **N**
9. 50 − 49 = **1**   **A**
10. 72 − 52 = **20**   **T**
11. 36 − 18 = **18**   **R**
12. 40 − 31 = **9**   **I**
13. 24 − 8 = **16**   **P**

▶ **Mixed Review**

Add.

14. 71 + 19 = **90**   27 + 27 = **54**   14 + 23 = **37**

15. 25 + 26 = **51**   7 + 18 = **25**   37 + 30 = **67**

PW60 Practice/Homework

Name _____

LESSON 10.5

Practice 2-Digit Subtraction

Circle the exercises in which you will need to regroup. Then subtract.

1. 67
   −32
   ‾‾
   35

2. 34
   −13
   ‾‾
   21

3. (⁵¹²
    6̸2̸
   −48
   ‾‾
    14)

4. (³¹⁴
    4̸4̸
   −16
   ‾‾
    28)

5. 76
   −40
   ‾‾
    36

6. (⁵¹¹
    6̸1̸
   −28
   ‾‾
    33)

7. 79
   −18
   ‾‾
    61

8. (³¹³
    4̸3̸
   −29
   ‾‾
    14)

9. (⁵¹⁵
    6̸5̸
   −38
   ‾‾
    27)

10. 95
   −42
   ‾‾
    53

11. 81
   −50
   ‾‾
    31

12. 56
   −15
   ‾‾
    41

13. (⁶¹⁴
    7̸4̸
   −36
   ‾‾
    38)

14. (⁷¹²
    8̸2̸
   −55
   ‾‾
    27)

15. (⁵¹⁰
    6̸0̸
   −26
   ‾‾
    34)

16. (⁴¹⁴
    5̸4̸
   −26
   ‾‾
    28)

17. 86
   −43
   ‾‾
    43

18. (⁸¹²
    9̸2̸
   −65
   ‾‾
    27)

19. (³¹⁰
    4̸0̸
   −27
   ‾‾
    13)

20. (⁵¹⁷
    6̸7̸
   −19
   ‾‾
    48)

 Mixed Review

Draw a line through any information you do not need. Then solve. Show your work.

21. Zoe has 14 pennies, 19 dimes, and ~~3 dollar bills~~ to spend at the store. How many coins does Zoe have?

¹
 14
+19
‾‾
 33

Practice/Homework   PW61

Mixed Practice

Circle the **+** or the **−**. Then solve.

1. 95 ⊖ 32 = 63
2. 52 ⊕ 27 = 79
3. ⁵¹⁷ 6̶7̶ ⊖ 8 = 59
4. ⁶¹⁸ 7̶8̶ ⊖ 59 = 19
5. ⁷¹⁶ 8̶6̶ ⊖ 18 = 68

6. 75 ⊕ 24 = 99
7. ¹25 ⊕ 36 = 61
8. ⁸¹⁴ 9̶4̶ ⊖ 48 = 46
9. ¹46 ⊕ 24 = 70
10. 50 ⊕ 38 = 88

11. 74 ⊖ 12 = 62
12. ⁴¹² 5̶2̶ ⊖ 49 = 3
13. 89 ⊖ 15 = 74
14. ¹44 ⊕ 37 = 81
15. ⁵¹² 6̶2̶ ⊖ 55 = 7

16. 22 ⊕ 77 = 99
17. ¹27 ⊕ 25 = 52
18. 42 ⊕ 37 = 79
19. ⁵¹⁰ 6̶0̶ ⊖ 41 = 19
20. ⁴¹⁶ 5̶6̶ ⊖ 18 = 38

▶ Mixed Review

Find the sum or difference.

21. $8 + 7 = \underline{15}$
22. $7 + 8 = \underline{15}$
23. $15 - 8 = \underline{7}$
24. $15 - 7 = \underline{8}$
25. $7 + 6 = \underline{13}$
26. $6 + 7 = \underline{13}$
27. $13 - 6 = \underline{7}$
28. $13 - 7 = \underline{6}$
29. $9 + 7 = \underline{16}$
30. $7 + 9 = \underline{16}$
31. $16 - 7 = \underline{9}$
32. $16 - 9 = \underline{7}$

Practice/Homework

Name _____

LESSON 10.7

Problem Solving • Too Much Information

Draw a line through any information you do not need. Then solve.

1. Robbie found 23 seashells at the beach. ~~18 of the seashells were pink.~~ He gave his sister 14 seashells. How many seashells does Robbie have left?

   $$\begin{array}{r} \overset{1\ 13}{\cancel{23}} \\ -14 \\ \hline 9 \end{array}$$

   __9__ seashells

2. Tracy grew 45 flowers ~~and 34 tomato plants~~ in her garden. She picked 17 flowers. How many flowers does she have left in her garden?

   $$\begin{array}{r} \overset{3\ 15}{\cancel{45}} \\ -17 \\ \hline 28 \end{array}$$

   __28__ flowers

3. Mr. Wesley's class made 38 paper frogs. ~~One half of the frogs were made from blue paper.~~ Mr. Wesley hung 26 of the frogs on the wall. How many frogs did Mr. Wesley **not** hang on the wall?

   $$\begin{array}{r} \overset{2\ 18}{\cancel{38}} \\ -26 \\ \hline 12 \end{array}$$

   __12__ frogs

4. ~~There are 18 girls on Kim's soccer team.~~ Kim's coach brought 15 soccer balls to practice. The team used 12 of the balls. How many soccer balls did the team **not** use?

   $$\begin{array}{r} 15 \\ -12 \\ \hline 3 \end{array}$$

   __3__ soccer balls

Practice/Homework

Name _____

LESSON 11.1

Pennies, Nickels, Dimes, and Quarters

Count on to find the total amount.
Circle the amount that can buy the toy.

57¢

1.

   <u>10</u>¢, <u>20</u>¢, <u>25</u>¢, <u>30</u>¢, <u>35</u>¢, <u>36</u>¢ | 36 | ¢

2.

   <u>25</u>¢, <u>35</u>¢, <u>40</u>¢, <u>45</u>¢ | 45 | ¢

3.

   <u>25</u>¢, <u>35</u>¢, <u>36</u>¢, <u>37</u>¢, <u>38</u>¢ | 38 | ¢

4.

   <u>25</u>¢, <u>35</u>¢, <u>45</u>¢, <u>55</u>¢, <u>56</u>¢, <u>57</u>¢   (57) ¢

▶ **Mixed Review**

Complete.

5. 12, <u>15</u>, 18, 21   25, 30, <u>35</u>, 40

6. <u>40</u>, 50, 60, 70   32, <u>34</u>, 36, 38

PW64   Practice/Homework

Name _____ 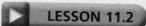   LESSON 11.2

Count Collections

Draw and label the coins needed to buy the item.
Put them in order from greatest to least value.
Write the total amount.   **Possible answers are shown.**

1.

   (25¢) (10¢) (10¢) (5¢)   __50__ ¢

2.

   (25¢) (5¢) (5¢) (5¢) (1¢)   __41__ ¢

3.

   (25¢) (25¢) (5¢) (1¢) (1¢) (1¢)   __58__ ¢

▶ **Mixed Review**

Write > or <.

4. 80 + 7 ⟨>⟩ 70 + 5 | 20 + 7 ⟨<⟩ 20 + 9 | 10 + 3 ⟨>⟩ 9 + 2
5. 60 + 1 ⟨<⟩ 60 + 9 | 40 + 7 ⟨>⟩ 40 + 2 | 10 + 4 ⟨>⟩ 8 + 5

Practice/Homework   PW65

Name _____

Make the Same Amounts

Use coins. Show the amount of money in two ways. **Check children's**
Draw and label each coin.   **drawings. Answers will vary.**

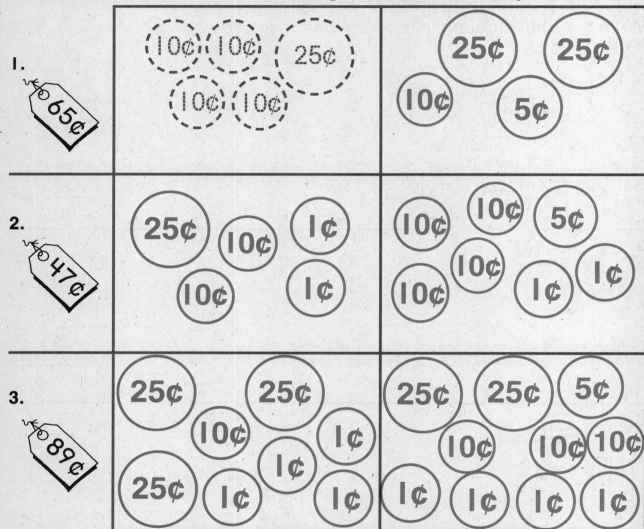

▶ **Mixed Review**

Solve.

4. 12 − 3 = __9__   9 + 5 = __14__   7 + 7 = __14__

5. 12 − 9 = __3__   16 − 8 = __8__   13 − 13 = __0__

6. 7 + 9 = __16__   9 + 9 = __18__   9 + 8 = __17__

PW66 Practice/Homework

Name _____

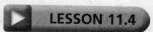

Algebra: Same Amounts Using the Fewest Coins

Write the amount. Then show the same amount with the fewest coins. Draw and label each coin. **Check children's drawings.**

Practice/Homework PW67

Name _____

LESSON 11.5

Make Change to $1.00

Count on from the price to find the change.

1. You have 55¢. You buy

__44__¢, __45__¢, __55__¢

Your change is __12¢__.

2. You have 50¢. You buy

__44__¢, __45__¢, __50__¢

Your change is __7¢__.

3. You have 70¢. You buy

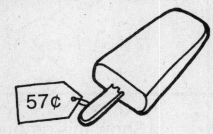

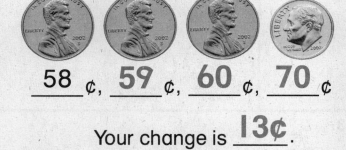

__58__¢, __59__¢, __60__¢, __70__¢

Your change is __13¢__.

▶ **Mixed Review**

Solve.

4. 7 + __8__ = 15   6 + __8__ = 14   6 + __6__ = 12

5. 16 − __8__ = 8   12 − __5__ = 7   14 − __6__ = 8

6. 9 + __6__ = 15   5 + __9__ = 14   9 + __9__ = 18

Name _____

LESSON 11.6

Count Bills and Coins

Count the bills and coins. Start with the bills.
Then count the coins in order from greatest to least value.
Write the total amount.

1.
   Total
   $5.00 , $6.00 , $6.10 , $6.15 $6.15

2.
   Total
   $5.00 , $5.10 , $5.11 , $5.12 $5.12

3.
   Total
   $5.00 , $10.00 , $10.25 , $10.30 $10.30

4.
   Total
   $5.00 , $6.00 , $6.10 , $6.11 $6.11

▶ **Mixed Review**

Find the sum or difference.

5. 4 + 8 = __12__   6. 7 + 6 = __13__   7. 8 + 7 = __15__

8. 15 − 7 = __8__   9. 12 − 7 = __5__   10. 14 − 6 = __8__

Name _____

> LESSON 11.7

More Bills and Coins

Count the bills and coins.
Write the total amount.

1.
    Total
$\underline{\$10.00}$, $\underline{\$11.00}$, $\underline{\$11.10}$, $\underline{\$11.20}$ $\underline{\$11.20}$

2.
     Total
$\underline{\$20.00}$, $\underline{\$21.00}$, $\underline{\$22.00}$, $\underline{\$22.05}$ $\underline{\$22.05}$

3.
     Total
$\underline{\$10.00}$, $\underline{\$15.00}$, $\underline{\$15.25}$, $\underline{\$15.35}$ $\underline{\$15.35}$

4.
      Total
$\underline{\$20.00}$, $\underline{\$25.00}$, $\underline{\$26.00}$, $\underline{\$26.25}$ $\underline{\$26.25}$

▶ **Mixed Review**

Add.

5. 4 + 6 = __10__   7 + 3 = __10__   2 + 8 = __10__

6. 9 + 3 = __12__   4 + 8 = __12__   6 + 6 = __12__

Name _____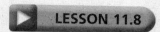

Problem Solving • Act It Out

Use bills and coins to act out the problems.
Then complete the tables. **Answers will vary.**

1. Denise wants to buy a book that costs $7.55. She has one $5 bill, seven $1 bills, three quarters, five dimes, and ten nickels. What are some different ways she can pay for the book?

$5 bill	$1 bills	quarters	dimes	nickels
1	2	2	0	1
1	2	1	3	0
0	7	2	0	1
0	6	3	5	10

2. Brett wants to buy a CD that costs $11.40. He has one $10 bill, two $5 bills, three $1 bills, six quarters, three dimes, and four nickels. What are some different ways Brett can pay for the CD?

$10 bill	$5 bills	$1 bills	quarters	dimes	nickels
1	0	1	1	1	1
0	2	1	0	3	2
0	1	0	4	3	2
0	2	1	1	0	3

Practice/Homework PW71

Name _____

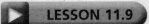

Make Change to $5.00

Count on from the price to find the change.

1. You have one $5 bill.
   You buy

$3.60, $3.70, $3.75, $4.00, $5.00

Your change is $1.40.

2. You have one $5 bill.
   You buy

$3.70, $3.75, $4.00, $5.00

Your change is $1.30.

3. You have one $5 bill.
   You buy

$4.30, $4.40, $4.50, $4.75, $5.00

Your change is $0.70.

▶ **Mixed Review**

Subtract.

4. 27 − 14 = __13__   75 − 10 = __65__   19 − 11 = __8__
5. 43 − 15 = __28__   36 − 18 = __18__   51 − 36 = __15__
6. 62 − 19 = __43__   30 − 12 = __18__   48 − 29 = __19__

PW72 Practice/Homework

Name _____   LESSON 11.10

Make Change to $10.00

Count on from the price to find the change.

1. You have one $10 bill.
   You buy

   $4.75, $5.00 , $10.00

   Your change is $5.25 .

2. You have one $10 bill.
   You buy

   $4.65, $4.75 , $5.00 , $10.00

   Your change is $5.35 .

3. You have one $10 bill.
   You buy

   $8.85, $8.90 , $9.00 , $10.00

   Your change is $1.15 .

▶ **Mixed Review**

Add.

4. 25   42   26   38   45
   +12   +27   +18   +24   +16
   ───   ───   ───   ───   ───
    37   69   44   62   61

Practice/Homework   PW73

Name _____

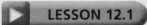 **LESSON 12.1**

Time to the Hour

Draw the minute hand and the hour hand.
Write the time another way.

1. 5 o'clock
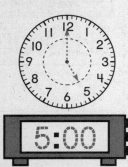

2. 8 o'clock

3. 10 o'clock

4. 3 o'clock

5. 11 o'clock

6. 2 o'clock

7. 6 o'clock

8. 9 o'clock

 Mixed Review

Write the value.

9.

___60___ ¢

10.

___16___ ¢

PW74 Practice/Homework

Name _____

LESSON 12.2

Time to the Half-Hour

Draw the minute hand.
Write the time another way.

1. 30 minutes after 5  **5:30**	2. eight-thirty  **8:30**	3. half past 6  **6:30**
4. 30 minutes before 2  **1:30**	5. four-thirty  **4:30**	6. 30 minutes after 10  **10:30**

▶ **Mixed Review**

Solve.

7. $25 - 10 =$ __15__ $48 + 12 =$ __60__ $36 + 8 =$ __44__

8. $71 - 23 =$ __48__ $42 + 33 =$ __75__ $50 - 26 =$ __24__

9. $14 + 17 =$ __31__ $75 - 32 =$ __43__ $57 + 15 =$ __72__

Practice/Homework PW75

Name _____

LESSON 12.3

Time to 15 Minutes

Draw the minute hand to show the time.
Write the time another way.

1. quarter past 2

2:15

2. eight-thirty

8:30

3. 5 o'clock

5:00

4. 45 minutes after 11

11:45

5. quarter to 6

5:45

6. half past 3

3:30

▶ **Mixed Review**

You have 75¢. Tell how much change you will get.

7.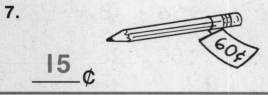

____15____ ¢

8.

____7____ ¢

9.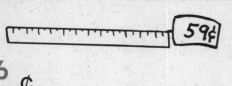

____16____ ¢

10.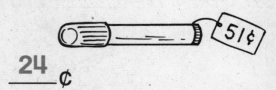

____24____ ¢

PW76   Practice/Homework

Name _____

LESSON 12.4

Time to 5 Minutes

Write the time another way.

1.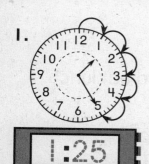
   1:25

2.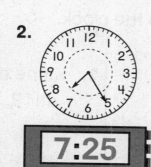
   7:25

3.
   10:35

4.
   8:40

5.
   11:20

6.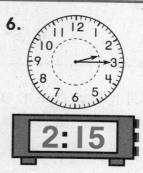
   2:15

7.
   5:50

8.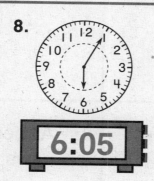
   6:05

9.
   12:35

▶ **Mixed Review**

Add.

10.
```
  15      23      72       9      10
  22      41      10      25      36
+ 31    + 17    +  8    + 38    + 17
  68      81      90      72      63
```

Practice/Homework **PW77**

Name _____

▶ **LESSON 12.5**

Problem Solving • Use a Model

Use a to show each time.
Then draw the minute hand on the clock.

1. Sam begins to play tennis at 3:30.

2. June begins to eat lunch at 12:10.

3. Bill takes a nap at 3:15.

4. Sue and her family leave to go to the beach at 9:15.

   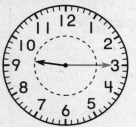

5. Allison begins to read her book at 4:00.

6. Ali begins delivering newspapers at 6:30.

PW78 Practice/Homework

Name _____   LESSON 12.6

Hours, Days, Weeks, Months, Years

Write *more than*, *less than*, or *the same as* to complete each sentence.

Time Relationships
There are 7 days in 1 week.
There are 28, 30, or 31 days in 1 month.
There are about 4 weeks in 1 month.
There are about 52 weeks in 1 year.
There are 12 months in 1 year.

1. Steve plays ball after school every day for 4 days in a row.

   This is ___less than___ 1 week.

2. Pam walks her dog every day for 1 month.

   This is ___less than___ 35 days.

3. Tim goes to summer camp for 45 days.

   This is ___more than___ 1 month.

4. Maya's family goes on vacation for 7 days.

   This is ___the same as___ 1 week.

5. The soccer season lasts 2 months.

   This is ___more than___ 7 weeks.

▶ **Mixed Review**

Write the time another way.

6.

Practice/Homework PW79

Name _____

LESSON 12.7

Sequence Events

This is Todd's evening schedule on school nights.

Todd's Evening Schedule		
Activity	Start Time	End Time
Eat dinner	6:00	6:30
Watch television	6:30	7:00
Take a shower	7:00	7:30
Read	7:30	8:00

Write **before** or **after** to complete the sentence.

1. Todd takes a shower ___**before**___ he reads.

2. Todd reads ___**after**___ he eats dinner.

3. Todd eats dinner ___**before**___ he takes a shower.

4. Todd watches television ___**before**___ he reads.

▶ **Mixed Review**

Solve.

5. $64 - 7 = \underline{57}$ $57 - 9 = \underline{48}$ $28 - 9 = \underline{19}$

6. $39 - 9 = \underline{30}$ $42 - 3 = \underline{39}$ $18 - 7 = \underline{11}$

Practice/Homework

Name _____

LESSON 13.1

Plane Shapes

Follow the directions. Check children's work.

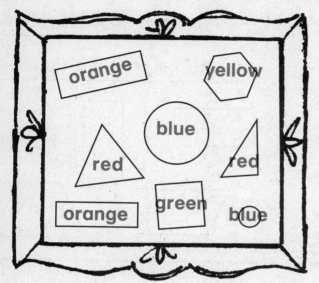

1. Color the rectangles orange.
2. Color the circles blue.
3. Color the hexagons yellow.
4. Color the triangles red.
5. Color the squares green.

6. Color the circles yellow.
7. Color the trapezoids red.
8. Color the hexagons green.
9. Color the parallelogram blue.
10. Color the pentagons orange.

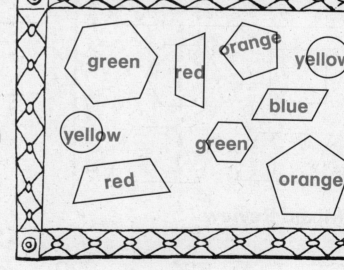

▶ **Mixed Review**

Subtract.

11. 43 − 5 = __38__   33 − 5 = __28__   18 − 5 = __13__
12. 27 − 5 = __22__   41 − 5 = __36__   94 − 5 = __89__
13. 45 − 5 = __40__   64 − 5 = __59__   70 − 5 = __65__

Practice/Homework PW81

Name _____

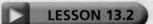

Algebra: Sort Plane Shapes

Answers may vary. Possible answers are given.

Write a title for each group of plane shapes.

1. __Shapes with 6 sides,__
   __6 angles, or 6 vertices__

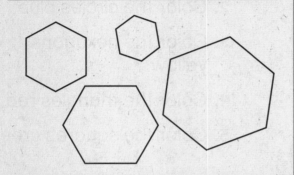

2. __Shapes with 4 sides,__
   __4 angles, or 4 vertices__

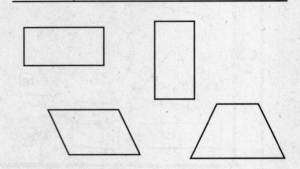

3. __Shapes with more than__
   __3 sides__

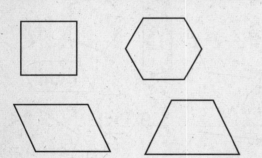

4. __Shapes with__
   __square corners__

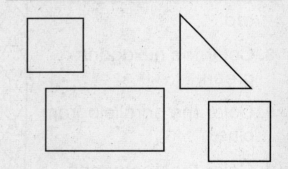

▶ **Mixed Review**

Add.

5. 37   6. 53   7. 20   8. 14   9. 33
    30   28   32   17   19
    + 4   +10   +46   +25   +27
    ―――   ―――   ―――   ―――   ―――
    71   91   98   56   79

Name _____

LESSON 13.3

Plane Shapes with 4 Sides

Follow the directions to color the design.

1. Color the quadrilaterals [Red].
2. Color the shapes that are not quadrilaterals [Blue].
   Check children's work. The two triangles and the pentagon should be blue. All other shapes should be red.

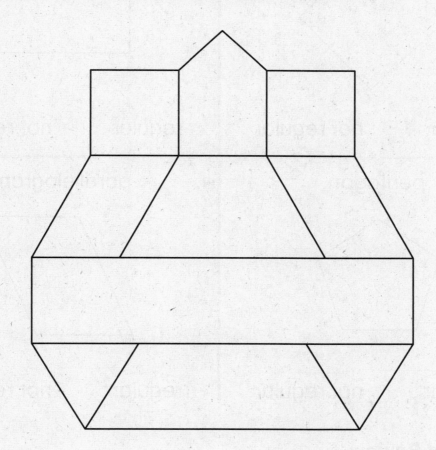

▶ **Mixed Review**

Find the sum.

3. 76 + 2 = __78__
4. 38 + 2 = __40__
5. 46 + 4 = __50__
6. 25 + 4 = __29__
7. 15 + 6 = __21__
8. 37 + 6 = __43__

Practice/Homework PW83

LESSON 13.4

Name _____

More About Plane Shapes

Tell whether the shape is regular or not regular.
Circle the answer.

1. square

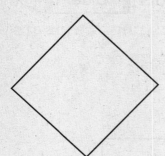

(regular)   not regular

2. rectangle

regular   (not regular)

3. pentagon

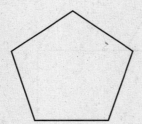

(regular)   not regular

4. parallelogram

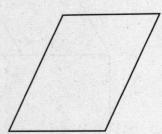

(regular)   not regular

▶ **Mixed Review**

Find the sum or difference.

5. $52 - 12 =$ __40__
6. $29 + 11 =$ __40__
7. $39 - 4 =$ __35__
8. $24 + 6 =$ __30__
9. $35 - 10 =$ __25__
10. $27 + 8 =$ __35__

PW84   Practice/Homework

Name _____

LESSON 13.5

Angles of Plane Shapes

Circle the shapes that belong.

1. shapes with **right angles**

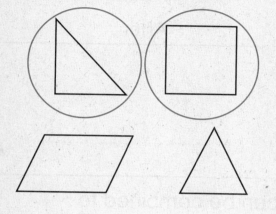

2. shapes with **obtuse angles**

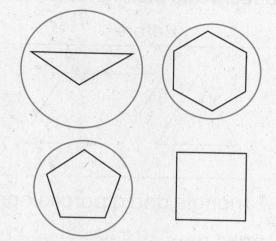

3. shapes with **no right angles**

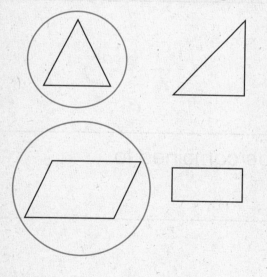

4. shapes with **acute angles**

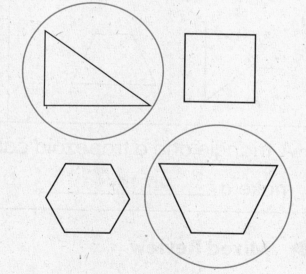

▶ **Mixed Review**

Write <, >, or =.

5. 4 + 2 ⊙> 3 + 1   6 + 5 ⊙= 7 + 4   8 + 4 ⊙> 9 + 2

6. 10 + 2 ⊙< 3 + 20   4 + 1 ⊙> 2 + 2   19 + 5 ⊙= 15 + 9

Practice/Homework **PW85**

Name _____

LESSON 13.6

Combine and Separate Shapes

Use pattern blocks or paper shapes. Put shapes together to make new shapes. Draw the new shapes. Write their names.

	Before	After
1.	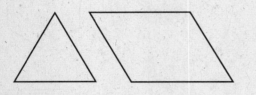	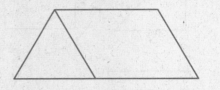

A triangle and a parallelogram can be combined to make a ___**trapezoid**___.

| 2. | 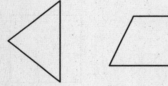 | 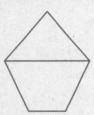 |

A triangle and a trapezoid can be combined to make a ___**pentagon**___.

 Mixed Review

3. Color the trapezoids green. Color the parallelograms red.

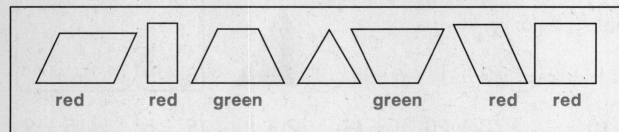

Name _____

LESSON 13.7

Problem Solving • Use Objects

Use objects. Make and draw a model with pattern blocks to solve.

1. Raquel has a hexagon. She wants to cut it into parallelograms. How many parallelograms can she cut from a hexagon?

   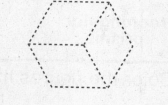

   __3__ parallelograms

2. Bridget has a parallelogram. She wants to divide it into triangles. How many triangles can she cut from the parallelogram?

   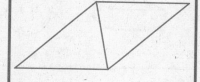

   __2__ triangles

3. Dylan cut a trapezoid into 3 equal parts. What pattern block shapes did he make?

    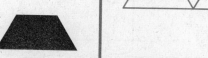

   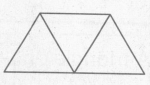

   __3 triangles__

4. Todd puts 6 triangles together by placing the points inward. What shape did he make?

   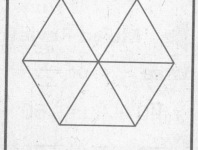

   __1 hexagon__

Practice/Homework PW87

Name _____

Solid Figures

     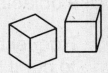

rectangular prisms   sphere   cones   cylinders   cubes

Color the figures that are the same shape.

1.

2.

3.

4.

5.

6.

▶ **Mixed Review**

Write >, <, or =.

7. $40 + 4 \;(<)\; 50 + 4$ $80 + 2 \;(>)\; 20 + 8$ $20 + 1 \;(=)\; 21 + 0$

8. $70 + 7 \;(=)\; 70 + 7$ $20 + 9 \;(<)\; 90 + 2$ $40 + 1 \;(>)\; 10 + 4$

9. $10 + 0 \;(>)\; 7 + 0$ $30 + 3 \;(>)\; 30 + 1$ $10 + 9 \;(=)\; 19 + 0$

PW88   Practice/Homework

Name _____

LESSON 14.2

Algebra: Sort Solid Figures

Complete the table. Write how many.

	Number of Faces, Edges, and Vertices		
solid figure	faces	edges	vertices
1. rectangular prism	6	12	8
2. cube	6	12	8
3. sphere	0	0	0

▶ **Mixed Review**

Count on to find the total amount.

4.

  25 ¢,　35 ¢,　45 ¢,　46 ¢,　47 ¢　| 47 | ¢

Name _____

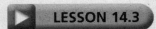

Compare Solid Figures and Plane Shapes

Use solid figures. Look at the faces.
Circle the solid figure you can make from the plane shapes.

1.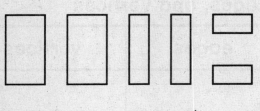

2.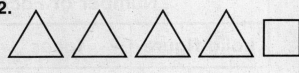

3.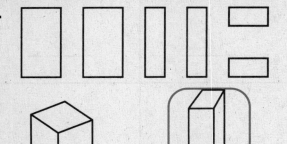

4.

▶ **Mixed Review**

Subtract.

5. $^{1}\cancel{2}^{17}$ -8 19	6. $^{4}\cancel{5}^{14}$ -9 45	7. $^{2}\cancel{3}^{17}$ -8 29	8. $^{2}\cancel{3}^{12}$ -6 26	9. $^{4}\cancel{5}^{11}$ -7 44
10. $^{5}\cancel{6}^{13}$ -8 55	11. $^{3}\cancel{4}^{14}$ -5 39	12. $^{7}\cancel{8}^{16}$ -8 78	13. $^{8}\cancel{9}^{15}$ -7 88	14. 24 -4 20

PW90   Practice/Homework

Name _____

LESSON 14.4

Combine Solid Figures

Circle the solid figures that make each solid object.

1.

2.

3.

4.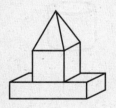

▶ **Mixed Review**

Find the sum.

5. 53   6. 45   7.   9   8. 37   9. 28
   + 7   +15   +13   +17   +14
   ----   ----   ----   ----   ----
    60   60   22   54   42

Practice/Homework PW91

Name _____

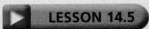

Take Apart Solid Figures

Name the solid figure. Tell how many faces, edges, and vertices it has. Then name the new solid figures. Tell how many faces, edges, and vertices they have in all.

1.

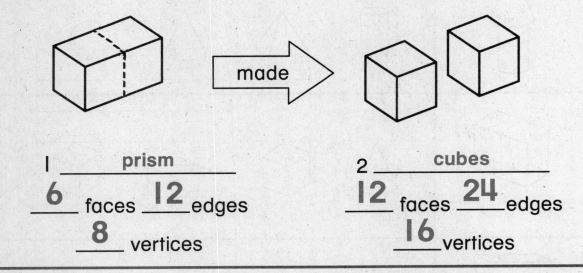

1 __prism__   2 __cubes__
__6__ faces __12__ edges   __12__ faces __24__ edges
__8__ vertices   __16__ vertices

Name the solid figure. Tell how many faces it has. Then name the new solid figures. Tell how many faces they have in all.

2.

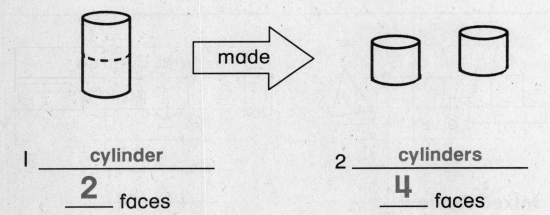

1 __cylinder__   2 __cylinders__
__2__ faces   __4__ faces

 Mixed Review

Find the difference.

3. 62 − 30 = __32__   4. 55 − 40 = __15__   5. 89 − 70 = __19__

Name _____

LESSON 14.6

Problem Solving • Use a Table

Complete the table and solve.

1. Jacob is making two solid figures. He plans to make 2 cubes. How many of each plane shape does he need to make the faces of the two figures?

Number of Shapes Needed for Solid Figures		
solid figure	rectangles	squares
cube	0	6
cube	0	6
total	0	12

Jacob needs __12__ squares.

---

2. Shannon is making four solid figures. She plans to make two rectangular prisms and two cubes. How many of each plane shape does she need to make the faces of the four figures? **Possible answers are given.**

Number of Shapes Needed for Solid Figures		
solid figure	other rectangles	squares
rectangular prism	6	0
rectangular prism	6	0
cube	0	6
cube	0	6
total	12	12

Shannon needs __12__ squares and __12__ other rectangles.

Practice/Homework PW93

Name _____

▶ LESSON 15.1

Length

Use your inch ruler.
Measure the picture.

1. 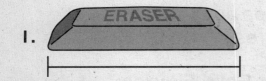 about __2__ inches

2. 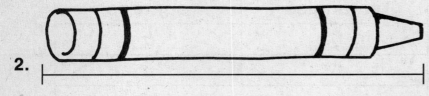 about __4__ inches

3. 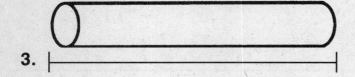 about __3__ inches

4. 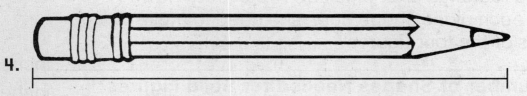 about __5__ inches

▶ **Mixed Review**

Write the number of faces on each solid figure.

5.   6.   7.
   __6__   __5__   __6__

PW94 Practice/Homework

Name _____

LESSON 15.2

Measure to the Nearest Inch

Work with a partner. Estimate the length.
Then measure to the nearest inch.   **Answers will vary.**

1. crayon

   Estimate: about _____ inches
   Measure: about _____ inches

2. book

   Estimate: about _____ inches
   Measure: about _____ inches

3. pencil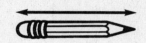

   Estimate: about _____ inches
   Measure: about _____ inches

4. tape

   Estimate: about _____ inches
   Measure: about _____ inches

5. stapler

   Estimate: about _____ inches
   Measure: about _____ inches

6. sheet of paper

   Estimate: about _____ inches
   Measure: about _____ inches

 Mixed Review

Add or subtract.

7.
```
  ¹              ¹            ² ¹⁵         ³ ¹²
   12            27            3̶5̶           4̶2̶           21
  + 9           + 5           − 7          − 9          +12
  ―――           ―――           ―――          ―――          ―――
   21            32            28           33           33
```

8.
```
   58            62            43           64           72
  −23           −10           +21          +31          −41
  ―――           ―――           ―――          ―――          ―――
   35            52            64           95           31
```

Practice/Homework   PW95

Name _____

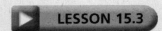

Inch, Foot, and Yard

Choose the best unit of measurement. **Check children's work.**
Write the measurement in inches, feet, or yards.

	Object	Unit	Measurement
1.	scissors	_____	about _____
2.	chalkboard	_____	about _____
3.	desk to door	_____	about _____
4.	door	_____	about _____
5.	stapler	_____	about _____

▶ **Mixed Review**

Write the time another way.

6.

7.

PW 96   Practice/Homework

Name _____

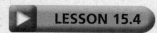

Inches and Feet

Measure the same object twice. Measure to the nearest inch.
Then measure to the nearest foot. Answers will vary.
Check children's work.

Find the object.	Measure in inches.	Measure in feet.
1. globe	about _____ inches	about _____ feet
2. window	about _____ inches	about _____ feet
3. chair	about _____ inches	about _____ feet

▶ **Mixed Review**

Write the number of vertices on each solid figure.

7.    8.   9.

___8___ vertices   ___0___ vertices   ___8___ vertices

Practice/Homework PW97

Name _____

Feet and Yards

Measure the same object twice.
Measure to the nearest foot.
Then measure to the nearest yard.

LESSON 15.5

Answers will vary.
Check children's work.

	Find the object.	Measure in feet.	Measure in yards.
1.	bookcase	about _____ feet	about _____ yards
2.	door	about _____ feet	about _____ yards
3.	oven	about _____ feet	about _____ yards

▶ **Mixed Review**

Find the sum.

4. 27   5. 18   6. 53   7. 34   8. 29
   +10   +11   + 5   +17   + 8
   ---   ---   ---   ---   ---
    37   29   58   51   37

PW98 Practice/Homework

Name _____

LESSON 15.6

Fahrenheit Temperature

Read the thermometer. Write the temperature.

1.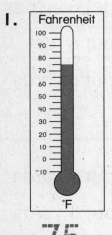
75 °F

2.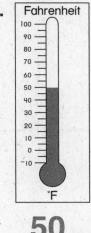
50 °F

3.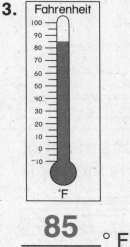
85 °F

4.
35 °F

5.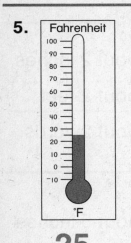
25 °F

6.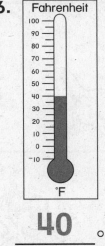
40 °F

7.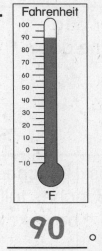
90 °F

8.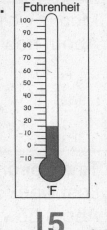
15 °F

▶ **Mixed Review**

Subtract.

9. 29
   − 4
   ───
    25

10. 33
    − 8
    ───
     25

11. 49
    − 9
    ───
     40

12. 61
    − 6
    ───
     55

13. 92
    − 6
    ───
     86

Practice/Homework PW99

Name _____

LESSON 15.7

Problem Solving • Make Reasonable Estimates

Circle the best estimate.

1. Bill is resting on the couch.
   About how long is the couch?

   about 6 inches
   (about 6 feet)
   about 6 yards

2. Nora swims across the pool.
   About how wide is the pool?

   about 2 yards
   about 5 yards
   (about 11 yards)

3. Mari is walking her dog.
   About how tall is the dog?

   about 2 inches
   (about 2 feet)
   about 2 yards

4. Tina's bean plant is 10 inches tall.
   Last week it was 8 inches tall.
   About how tall will it be next week?

   about 5 inches
   (about 13 inches)
   about 19 inches

5. Jacy is building a snowman.
   About what temperature
   is it outside?

   (about 20°F)
   about 50°F
   about 80°F

PW100 Practice/Homework

Name _____

LESSON 15.8

Problem Solving •
Choose the Measuring Tool and Unit

Measure a paper cup in different ways.
Then complete the chart. **Check children's work.**

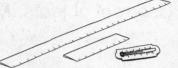

	what to measure	tool	measurement
1.	the cup's length	**ruler** _____	about ____ inches
2.	the cup's height	**ruler** _____	about ____ inches
3.	the temperature of the water	**thermometer** _____	about ____ °F

Measure a pitcher in different ways.
Then complete the chart. **Check children's work.**

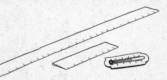

	what to measure	tool	measurement
4.	the pitcher's length	**ruler** _____	about ____ inches
5.	the pitcher's height	**ruler** _____	about ____ inches
6.	the temperature of the water	**thermometer** _____	about ____ °F

Practice/Homework PW101

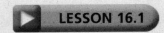

Name _____

Measure to the Nearest Centimeter

Use real objects. Measure to the nearest centimeter.
Answers will vary. Check children's work.

1. crayon	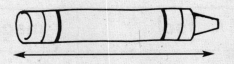	about _____ centimeters
2. milk carton		about _____ centimeters
3. glue stick		about _____ centimeters
4. tape	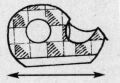	about _____ centimeters
5. safety pin	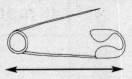	about _____ centimeters

▶ **Mixed Review**

Write the time another way.

6.

7.

PW102 Practice/Homework

Name _____

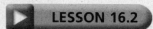 **LESSON 16.2**

Explore Centimeters and Meters

Measure each object in centimeters or meters. **Answers will vary. Check children's work.**

	Find the object.	Measure in centimeters.	Measure in meters.
1.	car	about _____ centimeters	about _____ meters
2.	refrigerator	about _____ centimeters	about _____ meters
3.	kitchen	about _____ centimeters	about _____ meters

 Mixed Review

Find the sum.

4. 28 + 16 = __44__   37 + 40 = __77__   52 + 29 = __81__
5. 45 + 5 = __50__   72 + 9 = __81__   64 + 7 = __71__
6. 13 + 16 = __29__   58 + 22 = __80__   23 + 49 = __72__

Practice/Homework   PW103

Name _____

Centimeter and Meter

Choose a unit to measure. Write your measurement to the nearest **centimeter** or **meter**. Answers will vary. Check children's work.

	Find the object.	Choose the unit.	Then measure.
1.	brush	_____	about _____
2.	bed	_____	about _____
3.	window	_____	about _____

 Mixed Review

Write the amount.

4. 　　　　　　　　　36¢ _____

5. 　$2.52 _____

PW104　Practice/Homework

Name _____

LESSON 16.4

Problem Solving • Guess and Check

Use real objects. Guess the length of
each object in centimeters.
Then use a centimeter ruler to check. **Check children's answers.**

1.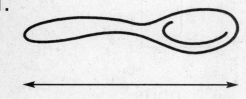

   Guess: _____ centimeters

   Check: about _____ centimeters

2.

   Guess: _____ centimeters

   Check: about _____ centimeters

3.

   Guess: _____ centimeters

   Check: about _____ centimeters

4.

   Guess: _____ centimeters

   Check: about _____ centimeters

Practice/Homework PW105

Explore Fractions

Write the number of parts.
Are the parts equal? Circle **yes** or **no**.

1.   (**yes**) / no
   __2__ parts

2.   yes / (**no**)
   __4__ parts

3.  yes / (**no**)
   __3__ parts

4. 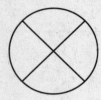 (**yes**) / no
   __4__ parts

5. 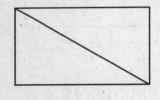 (**yes**) / no
   __2__ parts

6. yes / (**no**)
   __3__ parts

▶ Mixed Review

Write how many tens and ones. Write the number.

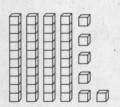

7. __4__ tens __6__ ones = __46__

8. __7__ tens __2__ ones = __72__

Name _____

LESSON 17.2

Unit Fractions

Draw lines to show equal parts. Color one part red.
Write the fraction for the red part.

1. 4 equal parts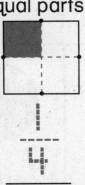
$\frac{1}{4}$

2. 3 equal parts
$\frac{1}{3}$

3. 8 equal parts
$\frac{1}{8}$

4. 6 equal parts
$\frac{1}{6}$

5. 4 equal parts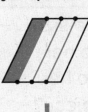
$\frac{1}{4}$

6. 2 equal parts
$\frac{1}{2}$

7. 3 equal parts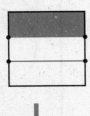
$\frac{1}{3}$

8. 2 equal parts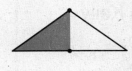
$\frac{1}{2}$

9. 10 equal parts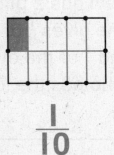
$\frac{1}{10}$

▶ **Mixed Review**

Solve.

10. $53 - 5 = \underline{48}$ $69 - 5 = \underline{64}$ $98 - 5 = \underline{93}$

11. $74 - 5 = \underline{69}$ $87 - 5 = \underline{82}$ $46 - 5 = \underline{41}$

Practice/Homework   PW107

Name _____

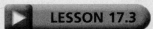
LESSON 17.3

Problem Solving • Use Objects

Use fraction bars or fraction circles.
Draw lines showing equal parts. Color one
part of each whole. Circle the fraction that is greater.
Answer the question. **Check children's work.**

1. Dan has 2 pancakes the same size. He gives $\frac{1}{3}$ of one pancake to Scott. He gives $\frac{1}{2}$ of the other to Ian. Who has the larger piece?

   Ian

   $\frac{1}{3}$ $\left(\frac{1}{2}\right)$

2. Tara has 2 waffles the same size. She gives $\frac{1}{4}$ of one waffle to her brother and $\frac{1}{3}$ of the other to her sister. Who has the larger piece?

   her sister

   $\frac{1}{4}$ $\left(\frac{1}{3}\right)$

3. Devi has 2 pitas the same size. She gives $\frac{1}{4}$ of one pita to Kelly and $\frac{1}{6}$ of the other to Jenny. Who has the larger piece?

   Kelly

   $\left(\frac{1}{4}\right)$ $\frac{1}{6}$

4. Joe has 2 granola bars the same size. He gives $\frac{1}{2}$ of one bar to Hector and $\frac{1}{3}$ of the other to Ryan. Who has the larger piece?

   Hector

   $\left(\frac{1}{2}\right)$ $\frac{1}{3}$

PW108   Practice/Homework

LESSON 17.4

Name _____

Other Fractions

Write the fraction for the shaded part.

1.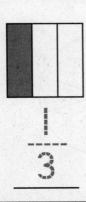
$$\frac{1}{3}$$

2.
$$\frac{2}{4}$$

3.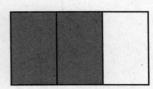
$$\frac{2}{3}$$

4.
$$\frac{5}{6}$$

5.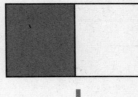
$$\frac{1}{2}$$

6.
$$\frac{3}{6}$$

7.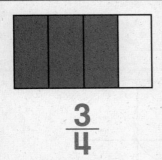
$$\frac{3}{4}$$

8.
$$\frac{1}{3}$$

9.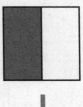
$$\frac{1}{2}$$

▶ **Mixed Review**

Solve.

10. $19 - \underline{11} = 8$ $30 - \underline{5} = 25$ $17 - \underline{9} = 8$

11. $10 - \underline{3} = 7$ $12 - \underline{5} = 7$ $20 - \underline{10} = 10$

12. $18 - \underline{7} = 11$ $14 - \underline{8} = 6$ $11 - \underline{6} = 5$

Practice/Homework PW109

Name _____

▶ LESSON 17.5

Fractions Equal to 1

Count the parts. Write each fraction.
Write the fraction for the whole.

1.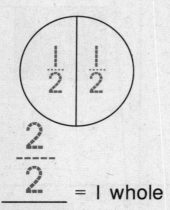

$\dfrac{2}{2}$ = 1 whole

2.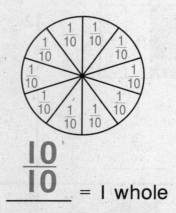

$\dfrac{10}{10}$ = 1 whole

3.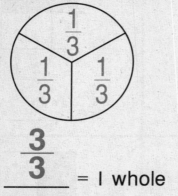

$\dfrac{3}{3}$ = 1 whole

4.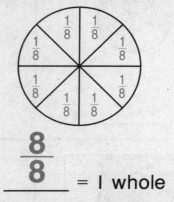

$\dfrac{8}{8}$ = 1 whole

5.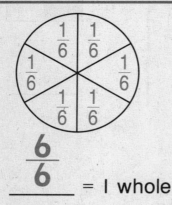

$\dfrac{6}{6}$ = 1 whole

6.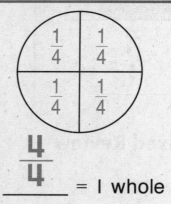

$\dfrac{4}{4}$ = 1 whole

▶ **Mixed Review**

Circle the better estimate.

7. a summer day   20°F   (80°F)

8. a winter day   (30°F)   70°F

PW110   Practice/Homework

Name _____

LESSON 17.6

Unit Fractions of a Set

Draw the set. Color one part red. **Check children's work.**
Write the fraction of the set for the red part.

1. 2 fish

   $\dfrac{1}{2}$

2. 6 cups

   $\dfrac{1}{6}$

3. 8 tables

   $\dfrac{1}{8}$

4. 10 balloons

   $\dfrac{1}{10}$

5. 4 boxes

   $\dfrac{1}{4}$

6. 3 buttons

   $\dfrac{1}{3}$

 Mixed Review

Write the temperature.

7. 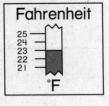 **23°F**

8. 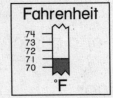 **71°F**

Practice/Homework   PW111

Name _____

LESSON 17.7

Other Fractions of a Set

Toss 4 two-color counters. Color the counters below to show how they land.

Answers will vary.

Write the fraction of the set for each color.

Sample answers are shown.

1.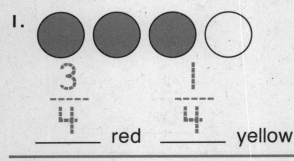

$\frac{3}{4}$ red $\frac{1}{4}$ yellow

2.

$\frac{4}{4}$ red 0 yellow

3.

$\frac{1}{4}$ red $\frac{3}{4}$ yellow

4.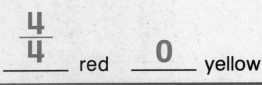

$\frac{2}{4}$ red $\frac{2}{4}$ yellow

5.

0 red $\frac{4}{4}$ yellow

6.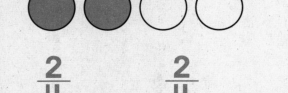

$\frac{3}{4}$ red $\frac{1}{4}$ yellow

▶ **Mixed Review**

Which measuring tool would you use to measure?
Write **ruler**, **yardstick**, or **thermometer**.

7. the length of a pencil _____ **ruler** _____

8. the temperature of a cup of water _____ **thermometer** _____

9. the height of a door _____ **yardstick** _____

PW112   Practice/Homework

Name _____

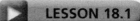 LESSON 18.1

Hundreds

Write how many hundreds, tens, and ones.

1.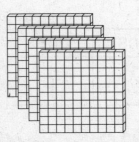

   __4__ hundreds
   __40__ tens
   __400__ ones

2.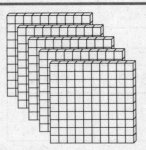

   __5__ hundreds
   __50__ tens
   __500__ ones

3.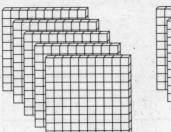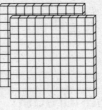

   __7__ hundreds
   __70__ tens
   __700__ ones

4.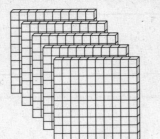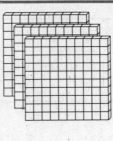

   __8__ hundreds
   __80__ tens
   __800__ ones

▶ **Mixed Review**

Solve.

5. 60 + 34 = __94__   44 + 52 = __96__   61 + 23 = __84__
6. 13 + 73 = __86__   40 + 18 = __58__   25 + 31 = __56__

Practice/Homework   PW113

Name _____

Hundreds, Tens, and Ones

Write the number in two different ways.

1.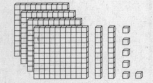

   four hundred twenty-six

hundreds	tens	ones
4	2	6

   400 + 20 + 6

2.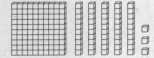

   one hundred fifty-three

hundreds	tens	ones
1	5	3

   100 + 50 + 3

3.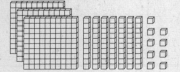

   three hundred seventy-nine

hundreds	tens	ones
3	7	9

   300 + 70 + 9

▶ **Mixed Review**

Add or subtract.

4. 72 − 51 = __21__   53 − 42 = __11__   66 − 50 = __16__

5. 12 + 9 = __21__   15 + 7 = __22__   18 + 6 = __24__

6. 57 − 24 = __33__   89 − 15 = __74__   64 − 33 = __31__

PW114 Practice/Homework

Name _____

Place Value

Circle the value of the underlined digit.

1. 3<u>6</u>4	2. <u>7</u>01	3. 25<u>9</u>
600 (60) 6	(700) 70 7	900 90 (9)
4. 54<u>8</u>	5. 4<u>6</u>3	6. 1<u>7</u>2
800 80 (8)	600 (60) 6	700 (70) 7
7. <u>6</u>07	8. 91<u>4</u>	9. <u>8</u>30
(600) 60 6	400 40 (4)	(800) 80 8
10. 52<u>6</u>	11. 1<u>8</u>1	12. <u>3</u>95
600 60 (6)	800 (80) 8	(300) 30 3
13. 4<u>3</u>7	14. <u>7</u>56	15. 40<u>1</u>
300 (30) 3	(700) 70 7	100 10 (1)

▶ **Mixed Review**

Write the missing numbers.

16. **51**, 52, 53   17, **18**, 19   63, **64**, 65

17. 34, 35, **36**   97, **98**, 99   **83**, 84, 85

Name _____

LESSON 18.4

Algebra: Different Ways to Show Numbers

Circle the correct ways to show each number.
Cross out and correct the other ways.

1.
| 928 |
| hundred | tens | ones |
| 9 | 2 | 8 |

900 + 80 + 2
~~20 + 8~~

(nine hundred twenty-eight)

~~2~~ hundreds ~~9~~ tens 8 ones
9   2

2.
117

~~700~~ + 10 + ~~X~~
100   7

(1 hundred 1 ten 7 ones)

one hundred seventy-~~one~~
            seventeen

3.
350

300 + ~~10~~ + ~~5~~   50

| hundred | tens | ones |
| 3 | 5 | 0 |

(three hundred fifty)

4.
604

600 + ~~40~~   4

(six hundred four)

(6 hundreds 0 tens 4 ones)

▶ **Mixed Review**

Write the fraction of the set for the shaded part.

5.

$\frac{2}{3}$

6.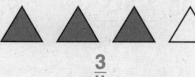

$\frac{3}{4}$

PW116 Practice/Homework

LESSON 18.5

Name _____

Problem Solving • Use Objects

Use dollar bills and coins. Count on.
Write the total amount.

1. Keb has two dollar bills, four dimes, and seven pennies. How much money does he have? $2.47 total

2. Sean has 1 dollar bill, six dimes, and eight pennies. How much money does he have? $1.68 total

3. Kwan has 6 dollar bills and five pennies. How much money does she have? $6.05 total

4. Devi has seven dollar bills, seven dimes, and two pennies. How much money does she have? $7.72 total

5. Beatrice has five dollar bills and nine dimes. How much money does she have? $5.90 total

6. Norm has four dollar bills, 1 dime, and six pennies. How much money does he have? $4.16 total

Practice/Homework PW117

Name _____

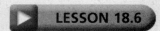

Explore Thousands

Write how many thousands and hundreds.

1. ___2___ thousands
   ___20___ hundreds

2. ___5___ thousands
   ___50___ hundreds

3. ___4___ thousands
   ___40___ hundreds

4. ___6___ thousands
   ___60___ hundreds

 Mixed Review

Write the time another way.

5.  1:15

6. 9:40

PW118   Practice/Homework

Name _____

▶ LESSON 18.7

Explore 4-Digit Numbers

Write the number in three different ways.

1.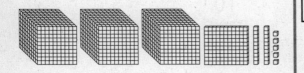

thousands	hundreds	tens	ones
3	1	2	5

3,000 + 100 + 20 + 5

3,125

2.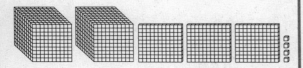

thousands	hundreds	tens	ones
2	3	0	4

2,000 + 300 + 0 + 4

2,304

3.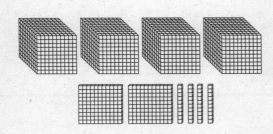

thousands	hundreds	tens	ones
4	2	4	0

4,000 + 200 + 40 + 0

4,240

▶ **Mixed Review**

Find each sum or difference.

4. 11 + 24 = __35__   52 − 29 = __23__   48 + 36 = __84__

5. 57 − 28 = __29__   64 + 31 = __95__   70 − 25 = __45__

6. 44 + 18 = __62__   52 − 20 = __32__   85 − 59 = __26__

Practice/Homework   PW119

Name _____

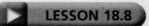

10, 100, 1,000

About how many objects are in each picture?
Circle the answer.

1.
   pencils

   (10)
   100
   1,000

2.
   people at a baseball game

   10
   100
   (1,000)

3.
   gumballs

   10
   (100)
   1,000

 Mixed Review

Write how many faces, edges, and vertices
on each solid figure.

4.
   cube

   __6__ faces
   __12__ edges
   __8__ vertices

5.
   sphere

   __0__ faces
   __0__ edges
   __0__ vertices

PW120 Practice/Homework

Name _____ 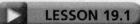 LESSON 19.1

Algebra: Use Symbols: >, <, and =

Compare.
Write >, <, or =.

1. $200 + 5 \; \boxed{<} \; 200 + 70 + 5$

2. $900 + 20 + 2 \; \boxed{<} \; 900 + 20 + 3$

3. $300 + 70 + 9 \; \boxed{>} \; 300 + 10 + 9$

4. $600 + 40 + 2 \; \boxed{>} \; 600 + 20 + 4$

5. $400 + 10 + 1 \; \boxed{=} \; 400 + 10 + 1$

6. $700 + 30 + 7 \; \boxed{=} \; 700 + 30 + 7$

7. $800 + 50 + 9 \; \boxed{<} \; 900 + 50 + 9$

8. $100 + 80 \; \boxed{>} \; 100 + 8$

▶ **Mixed Review**

Solve.

9. $14 + 81 = \underline{95}$ $44 + 44 = \underline{88}$ $8 + 61 = \underline{69}$

10. $53 - 5 = \underline{48}$ $77 - 22 = \underline{55}$ $97 - 30 = \underline{67}$

Name _____

Missing Numbers to 1,000

Use ▭ to show the numbers.
Write the missing numbers.

1. 205, **206**, 207, **208**, 209, 210, **211**, 212, **213**

2. 609, **610**, **611**, 612, 613, **614**, 615, **616**, 617

3. **527**, 526, **525**, 524, 523, **522**, 521, 520, **519**

4. **136**, **137**, 138, 139, 140, **141**, **142**, 143, 144

5. 311, **310**, 309, **308**, 307, **306**, 305, **304**, 303

6. **977**, 978, **979**, **980**, 981, 982, 983, **984**, 985

▶ **Mixed Review**
Write the total amount.

7.

27¢

8.

56¢

PW122 Practice/Homework

Name _____

Algebra: Skip-Count

Skip-count. Write the missing numbers.

1. 243, 253, 263, __273__, __283__, __293__, __303__, 313

2. 785, 790, 795, __800__, __805__, __810__, __815__, 820

3. 651, 655, 659, __663__, __667__, __671__, __675__, 679

4. 122, 124, 126, __128__, __130__, __132__, __134__, 136

5. 504, 507, 510, __513__, __516__, __519__, __522__, 525

6. 400, 450, 500, __550__, __600__, __650__, __700__, 750

▶ **Mixed Review**

Subtract.

7.
   39   63   42   53   77
− 18   − 24   − 35   − 28   − 40
  21   39   7   25   37

8.
   25   73   59   81   45
− 13   − 24   − 32   − 47   − 19
  12   49   27   34   26

Name _____

Problem Solving • Look For a Pattern

Write the pattern rule.
Write the next number in the pattern.

THINK: Look across the rows.

1.

401	402	403	404	405	406	407	408	409	410
411	412	413	414	415	416	417	418	419	420
421	422	423	424	425	426	427	428	429	430
431	432	433	434	435	436	437	438	439	440
441	442	443	444	445	446	447	448	449	450

Rule: __Count by 3s.__

Next number: __451__

2.

801	802	803	804	805	806	807	808	809	810
811	812	813	814	815	816	817	818	819	820
821	822	823	824	825	826	827	828	829	830
831	832	833	834	835	836	837	838	839	840
841	842	843	844	845	846	847	848	849	850

Rule: __Count by 2s.__

Next number: __852__

3.

101	102	103	104	105	106	107	108	109	110
111	112	113	114	115	116	117	118	119	120
121	122	123	124	125	126	127	128	129	130
131	132	133	134	135	136	137	138	139	140
141	142	143	144	145	146	147	148	149	150

Rule: __Count by 6s.__

Next number: __154__

Name _____

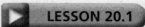 **LESSON 20.1**

Mental Math: Add Hundreds

Add. Write the missing numbers.

1.
   1 hundred   100
+ 4 hundreds   + 400
   5 hundreds   500

2.
   2 hundreds   200
+ 7 hundreds   + 700
   9 hundreds   900

3.
   3 hundreds   300
+ 5 hundreds   + 500
   8 hundreds   800

4.
   3 hundreds   300
+ 3 hundreds   + 300
   6 hundreds   600

5.
   4 hundreds   400
+ 0 hundreds   + 0
   4 hundreds   400

6.
   5 hundreds   500
+ 2 hundreds   + 200
   7 hundreds   700

▶ **Mixed Review**

7. 99 − 12 = __87__   68 − 41 = __27__   55 − 25 = __30__

8. 76 − 57 = __19__   47 − 32 = __15__   32 − 18 = __14__

9. 81 − 56 = __25__   27 − 18 = __9__   74 − 28 = __46__

Practice/Homework PW125

Name _____

Equal and Not Equal

Write = or ≠.

1. 100 + 300 ⊜ 400
2. 200 + 500 ⊜ 700
3. 700 + 200 ≠ 800
4. 400 + 300 ≠ 600
5. 300 + 500 ⊜ 800
6. 600 + 100 ≠ 500
7. 800 + 100 ⊜ 900
8. 500 + 400 ≠ 100
9. 100 + 600 ⊜ 700
10. 400 + 400 ⊜ 800
11. 200 + 100 ≠ 400
12. 700 + 100 ≠ 600

▶ **Mixed Review**

Solve.

13. 3 + 5 = __8__   5 + 4 = __9__   2 + 5 = __7__
14. 2 + 2 = __4__   2 + 7 = __9__   6 + 2 = __8__
15. 1 + 6 = __7__   3 + 4 = __7__   5 + 1 = __6__
16. 4 + 3 = __7__   1 + 8 = __9__   3 + 6 = __9__

PW126   Practice/Homework

Name _____

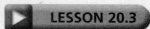

Model 3-Digit Addition: Regroup Ones

Use Workmat 5 and [flat] [rod] [unit]. Add.

1.
hundreds	tens	ones
	1	
2	3	9
+2	0	2
4	4	1

2.
hundreds	tens	ones
	1	
8	0	6
+1	2	7
9	3	3

3.
hundreds	tens	ones
	1	
1	2	9
+4	1	3
5	4	2

4.
hundreds	tens	ones
	1	
2	3	6
+3	1	6
5	5	2

5.
hundreds	tens	ones
	1	
8	0	7
+1	3	4
9	4	1

6.
hundreds	tens	ones
	1	
6	2	8
+	1	3
6	4	1

7.
hundreds	tens	ones
	1	
3	5	5
+2	1	8
5	7	3

8.
hundreds	tens	ones
	1	
8	1	8
+	7	3
8	9	1

9.
hundreds	tens	ones
	1	
5	4	7
+2	2	9
7	7	6

▶ **Mixed Review**

How many hundreds, tens, and ones are there?

10. 862 = __8__ hundreds __6__ tens __2__ ones

11. 729 = __7__ hundreds __2__ tens __9__ ones

12. 376 = __3__ hundreds __7__ tens __6__ ones

Practice/Homework   PW127

Name _____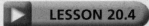

Model 3-Digit Addition: Regroup Tens

Use Workmat 5 and 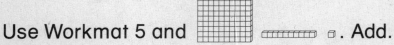. Add.

1.
hundreds	tens	ones
[1]	[1]	
2	1	7
+1	9	9
4	1	6

2.
hundreds	tens	ones
[1]	[1]	
4	2	9
+1	7	7
6	0	6

3.
hundreds	tens	ones
[1]	[1]	
2	2	5
+5	8	6
8	1	1

4. 429
  +187
  ───
  616

5. 675
  +153
  ───
  828

6. 321
  +296
  ───
  617

7. 523
  +406
  ───
  929

8. 199
  +730
  ───
  929

9. 462
  +450
  ───
  912

10. 610
  +198
  ───
  808

11. 725
  + 92
  ───
  817

▶ **Mixed Review**

Write the fraction for the shaded part.

12.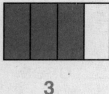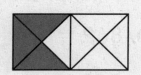

   $\dfrac{3}{4}$ $\dfrac{6}{8}$ $\dfrac{3}{8}$

Name _____

LESSON 20.5

More 3-Digit Addition

Add.

1. 234 + 117 **351**	2. 361 + 212 **573**	3. 460 + 295 **755**	4. 581 + 219 **800**
5. 332 + 208 **540**	6. 488 + 107 **595**	7. 367 + 323 **690**	8. 495 + 303 **798**
9. 627 + 192 **819**	10. 541 + 29 **570**	11. 681 + 248 **929**	12. 124 + 46 **170**
13. 520 + 50 **570**	14. 336 + 219 **555**	15. 475 + 16 **491**	16. 629 + 344 **973**

▶ **Mixed Review**

Write >, <, or =.

17. 15 + 2 ⟶ **>** ⟵ 12 + 4   17 + 3 ⟶ **=** ⟵ 12 + 8

18. 20 + 3 ⟶ **=** ⟵ 19 + 4   13 + 5 ⟶ **>** ⟵ 11 + 6

19. 18 + 2 ⟶ **=** ⟵ 13 + 7   12 + 9 ⟶ **<** ⟵ 18 + 4

Practice/Homework   PW129

Name _____

LESSON 20.6

Add Money

Add.

1. $3.45 + $2.28 = $5.73	2. $4.72 + $3.23 = $7.95	3. $5.61 + $0.95 = $6.56	4. $5.92 + $3.20 = $9.12
5. $4.43 + $4.19 = $8.62	6. $5.89 + $2.08 = $7.97	7. $4.78 + $4.40 = $9.18	8. $1.95 + $3.04 = $4.99
9. $7.38 + $2.33 = $9.71	10. $5.64 + $0.11 = $5.75	11. $4.92 + $3.54 = $8.46	12. $2.35 + $4.16 = $6.51
13. $6.30 + $1.98 = $8.28	14. $3.69 + $2.29 = $5.98	15. $4.84 + $1.25 = $6.09	16. $8.85 + $0.90 = $9.75

▶ **Mixed Review**

Are the parts equal? Circle **yes** or **no**.

17.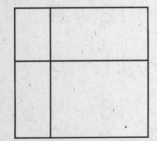

18.

PW130 Practice/Homework

Name _____

LESSON 20.7

Problem Solving • Choose the Computational Method

Choose a method to solve each problem.

1. Don has 230 baseball cards in one album and 242 in another. How many baseball cards does he have in all?

   __472__ cards

2. There are 165 first graders and 130 second graders at the dinosaur museum. How many children are at the museum altogether?

   __295__ children

3. Mr. Chen has collected 425 antique car magazines. He has collected 305 sports magazines. How many magazines has he collected in all?

   __730__ magazines

4. Craig has collected 556 bottle caps. Mary has collected 293 bottle caps. How many bottle caps have they both collected together?

   __849__ bottle caps

Practice/Homework PW131

Name _____

Mental Math: Subtract Hundreds

Subtract. Write the missing numbers.

1.
    5 hundreds   500
    − 1 hundred   − 100
    ―――――――――――
    4 hundreds   400

2.
    3 hundreds   300
    − 3 hundreds   − 300
    ―――――――――――
    0 hundreds   0

3.
    9 hundreds   900
    − 4 hundreds   − 400
    ―――――――――――
    5 hundreds   500

4.
    6 hundreds   600
    − 2 hundreds   − 200
    ―――――――――――
    4 hundreds   400

5.
    8 hundreds   800
    − 5 hundreds   − 500
    ―――――――――――
    3 hundreds   300

6.
    4 hundreds   400
    − 2 hundreds   − 200
    ―――――――――――
    2 hundreds   200

▶ **Mixed Review**

Add or subtract.

7. 72¢ − 11¢ = __61¢__   69¢ − 29¢ = __40¢__

8. 55¢ + 37¢ = __92¢__   42¢ + 33¢ = __75¢__

9. 86¢ − 49¢ = __37¢__   91¢ − 59¢ = __32¢__

Name _____

Model 3-Digit Subtraction: Regroup Tens

Use Workmat 5 and ▭ ▭ ▫. Subtract.

	H	T	O
1.	7	2̸/3̸	10̸/0̸
	− 4	1	2
	3	1	8

	H	T	O
2.	3	8̸/9̸	11̸/1̸
	− 2	0	4
	1	8	7

	H	T	O
3.	8	1̸/2̸	12̸/2̸
	− 1	0	6
	7	1	6

	H	T	O
4.	7	4̸/5̸	16̸/6̸
	− 2	4	8
	5	0	8

	H	T	O
5.	5	2̸/3̸	18̸/8̸
	− 1	1	9
	4	1	9

	H	T	O
6.	8	2̸/3̸	14̸/4̸
	−	2	7
	8	0	7

	H	T	O
7.	5	3̸/4̸	12̸/2̸
	− 2	1	8
	3	2	4

	H	T	O
8.	8	9	5
	− 4	7	3
	4	2	2

	H	T	O
9.	8	3̸/4̸	15̸/5̸
	− 3	2	9
	5	1	6

▶ **Mixed Review**

Solve.

10. 66 + 26 = __92__   28 − 18 = __10__
11. 78 − 28 = __50__   57 + 17 = __74__
12. 30 + 10 = __40__   84 − 34 = __50__

Name _____

Model 3-Digit Subtraction: Regroup Hundreds

Use Workmat 5 and ▭▭▭ ▫. Subtract.

1.	H	T	O
	⑤	⑩	
	6̸	0̸	6
−	2	5	2
	3	5	4

2.	H	T	O
	⑦	⑬	
	8̸	3̸	5
−	4	7	2
	3	6	3

3.	H	T	O
		⑤	⑫
	4	6̸	2̸
−		3	3
	4	2	9

4. ⁴¹⁰
   5̸0̸4
− 182
  322

5. ⁷¹²
   8̸2̸4
− 654
  170

6. ¹¹²
   2̸2̸9
− 86
  143

7. ⁷¹²
   6̸8̸2
− 663
   19

8. ²¹⁰
   3̸0̸3
− 111
  192

9. ⁸¹²
   9̸2̸4
− 193
  731

10. ³¹³
   5̸4̸3
− 527
   16

11. ⁷¹⁵
   6̸8̸5
− 478
  207

▶ **Mixed Review**

Write the number that is greater.

12. 916, 961   **961**
13. 777, 727   **777**
14. 227, 272   **272**
15. 111, 191   **191**
16. 585, 515   **585**
17. 629, 692   **692**

Name _____

**LESSON 21.4**

More 3-Digit Subtraction

Subtract.

1. $\overset{5\,13}{5\cancel{6}\cancel{3}}$ $-\,218$ $\overline{345}$	2. $\overset{3\,11}{\cancel{4}\cancel{1}9}$ $-\,236$ $\overline{183}$	3. 782 $-\,531$ $\overline{251}$	4. $\overset{8\,16}{29\cancel{6}}$ $-\,159$ $\overline{137}$
5. $\overset{7\,12}{1\cancel{8}2}$ $-\,65$ $\overline{117}$	6. $\overset{2\,14}{3\cancel{4}2}$ $-\,261$ $\overline{81}$	7. $\overset{8\,15}{69\cancel{5}}$ $-\,478$ $\overline{217}$	8. $\overset{4\,10}{5\cancel{0}3}$ $-\,232$ $\overline{271}$
9. $\overset{6\,10}{4\cancel{7}0}$ $-\,254$ $\overline{216}$	10. $\overset{1\,17}{82\cancel{7}}$ $-\,8$ $\overline{819}$	11. $\overset{0\,17}{7\cancel{1}\cancel{7}}$ $-\,309$ $\overline{408}$	12. 608 $-\,406$ $\overline{202}$
13. $\overset{1\,13}{2\cancel{3}5}$ $-\,95$ $\overline{140}$	14. $\overset{3\,14}{4\cancel{4}4}$ $-\,380$ $\overline{64}$	15. 629 $-\,222$ $\overline{407}$	16. $\overset{6\,16}{87\cancel{6}}$ $-\,729$ $\overline{147}$

▶ **Mixed Review**

Write the missing addends.

17. $5 + \underline{7} = 12 \qquad 8 + \underline{8} = 16 \qquad 3 + \underline{7} = 10$

18. $\underline{6} + 6 = 12 \qquad \underline{6} + 9 = 15 \qquad \underline{9} + 5 = 14$

19. $4 + \underline{9} = 13 \qquad \underline{5} + 4 = 9 \qquad \underline{8} + 3 = 11$

20. $3 + \underline{9} = 12 \qquad \underline{7} + 6 = 13 \qquad \underline{4} + 6 = 10$

Practice/Homework PW135

Name _____

 **LESSON 21.5**

Subtract Money

Subtract.

1. $\overset{6\;12}{\cancel{\$7}.\cancel{22}}$ $-\;\$3.50$ $\;\;\$3.72$	2. $\overset{3\;16}{\$5.\cancel{46}}$ $-\;\$2.37$ $\;\;\$3.09$	3. $\overset{6\;10}{\$6.\cancel{70}}$ $-\;\$1.08$ $\;\;\$5.62$	4. $\overset{8\;10}{\cancel{\$9}.09}$ $-\;\$5.79$ $\;\;\$3.30$
5. $\$5.34$ $-\;\$2.24$ $\;\;\$3.10$	6. $\$6.48$ $-\;\$2.27$ $\;\;\$4.21$	7. $\overset{2\;15}{\$7.\cancel{35}}$ $-\;\$6.09$ $\;\;\$1.26$	8. $\overset{5\;17}{\$4.\cancel{67}}$ $-\;\$0.38$ $\;\;\$4.29$
9. $\$3.86$ $-\;\$1.81$ $\;\;\$2.05$	10. $\overset{2\;15}{\$9.\cancel{45}}$ $-\;\$6.29$ $\;\;\$3.16$	11. $\overset{7\;11}{\$7.\cancel{81}}$ $-\;\$4.04$ $\;\;\$3.77$	12. $\overset{7\;15}{\$8.\cancel{52}}$ $-\;\$4.61$ $\;\;\$3.91$
13. $\overset{2\;10}{\cancel{\$8.30}}$ $-\;\$6.05$ $\;\;\$2.25$	14. $\overset{8\;10}{\cancel{\$3.90}}$ $-\;\$1.36$ $\;\;\$2.54$	15. $\$9.95$ $-\;\$4.50$ $\;\;\$5.45$	16. $\overset{6\;13}{\cancel{\$7}.30}$ $-\;\$4.60$ $\;\;\$2.70$

▶ **Mixed Review**

Write how many thousands and hundreds.

17.

__2__ hundreds = __20__ tens

18.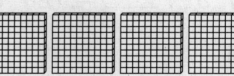

__4__ hundreds = __40__ tens

PW136 Practice/Homework

Name _____

LESSON 21.6

Practice 3-Digit Addition and Subtraction

Add or subtract to solve.

1. $\overset{0\,13}{4\cancel{1}3}$
    $-\;206$
    $\overline{207}$

2. $\overset{1}{577}$
    $+\;304$
    $\overline{881}$

3. 395
    $-\;212$
    $\overline{183}$

4. $\overset{1}{516}$
    $+\;392$
    $\overline{908}$

5. $\overset{1}{463}$
    $+\;228$
    $\overline{691}$

6. $\overset{1}{236}$
    $+\;119$
    $\overline{355}$

7. $\overset{6\,16}{7\cancel{6}4}$
    $-\;571$
    $\overline{193}$

8. $\overset{7\,15}{8\cancel{5}3}$
    $-\;382$
    $\overline{471}$

9. $\overset{1}{638}$
    $+\;171$
    $\overline{809}$

10. $\overset{2\,10}{6\cancel{3}\cancel{0}}$
    $-\;421$
    $\overline{209}$

11. 424
    $+\;22$
    $\overline{446}$

12. $\overset{4\,15}{5\cancel{5}5}$
    $-\;285$
    $\overline{270}$

13. $\overset{4\,12}{1\cancel{5}\cancel{2}}$
    $-\;38$
    $\overline{114}$

14. $\overset{1}{549}$
    $+\;328$
    $\overline{877}$

15. $\overset{1}{258}$
    $+\;132$
    $\overline{390}$

16. $\overset{5\,17}{6\cancel{7}9}$
    $-\;391$
    $\overline{288}$

  Mixed Review

Tell whether the shape is **regular** or **not regular**.

17. square

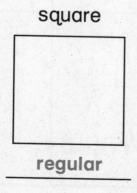

regular

18. trapezoid

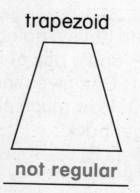

not regular

Name _____

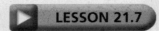
LESSON 21.7

Problem Solving • Solve Multistep Problems

Do one step at a time.
Add or subtract.

Steps may vary.

	Step 1	Step 2
1. Peter saves $7.45. His sister gives him $2.37. He uses the money to buy a toy car for $6.81. How much money does Peter have left? $3.01	$7.45 +$2.37 $9.82	$9.82 −$6.81 $3.01
2. Maria has $3.25 in her piggy bank. She earns $2.50 doing chores. She spends $2.10. How much money does Maria have left? $3.65	$3.25 +$2.50 $5.75	$5.75 −$2.10 $3.65
3. Wanda has $2.95. Her brother has $3.27. They need $8.15 to buy a present. How much more money do they need? $1.93	$2.95 +$3.27 $6.22	$8.15 −$6.22 $1.93
4. Juanita buys a notebook for $1.99 and a box of crayons for $2.01. She gives the clerk $5.00. How much change does she get back? $1.00	$1.99 +$2.01 $4.00	$5.00 −$4.00 $1.00

Practice/Homework

Name _____

Addition and Multiplication

Write the sum. Then write the product.

1.

   4 + 4 + 4 = __12__   3 × 4 = __12__

2. (cats images)

   3 + 3 + 3 + 3 = __12__   4 × 3 = __12__

3.

   5 + 5 + 5 = __15__   3 × 5 = __15__

4.

   1 + 1 + 1 + 1 + 1 = __5__   5 × 1 = __5__

▶ **Mixed Review**

Subtract.

5. 154 − 10 = __144__   149 − 10 = __139__   125 − 92 = __33__

6. 172 − 10 = __162__   138 − 26 = __112__   147 − 95 = __52__

7. 118 − 10 = __108__   194 − 61 = __133__   136 − 91 = __45__

Practice/Homework   PW139

Name _____

LESSON 22.2

Arrays

Write how many rows and how many in each row.
Multiply and write the product.

1.
   __4__ rows
   __5__ in each row

   $4 \times 5 = \underline{20}$

2.
   __3__ rows
   __6__ in each row

   $3 \times 6 = \underline{18}$

3.
   __2__ rows
   __8__ in each row

   $2 \times 8 = \underline{16}$

4.
   __6__ rows
   __1__ in each row

   $6 \times 1 = \underline{6}$

5.
   __5__ rows
   __5__ in each row

   $5 \times 5 = \underline{25}$

6.
   __6__ rows
   __4__ in each row

   $6 \times 4 = \underline{24}$

▶ **Mixed Review**

Write the number.

7. 3 hundreds, 4 tens, 7 ones __347__

8. 6 hundreds, 1 ten, 3 ones __613__

9. 5 hundreds, 5 tens, 1 one __551__

10. 8 hundreds, 3 tens, 2 ones __832__

PW140 Practice/Homework

Name _____

▶ LESSON 22.3

Algebra: Multiply in Any Order
Write the product.
Write the multiplication sentence in a different order.

1. $4 \times 5 = \underline{20}$

   $\underline{5} \times \underline{4} = \underline{20}$

2. $10 \times 3 = \underline{30}$

   $\underline{3} \times \underline{10} = \underline{30}$

3. $2 \times 9 = \underline{18}$

   $\underline{9} \times \underline{2} = \underline{18}$

4. $3 \times 7 = \underline{21}$

   $\underline{7} \times \underline{3} = \underline{21}$

5. $6 \times 3 = \underline{18}$

   $\underline{3} \times \underline{6} = \underline{18}$

6. $8 \times 2 = \underline{16}$

   $\underline{2} \times \underline{8} = \underline{16}$

7. $7 \times 10 = \underline{70}$

   $\underline{10} \times \underline{7} = \underline{70}$

8. $3 \times 8 = \underline{24}$

   $\underline{8} \times \underline{3} = \underline{24}$

▶ **Mixed Review**

Write the number that comes next.

9. 10, 20, 30, 40, __50__

10. 45, 55, 65, 75, __85__

11. 6, 12, 18, 24, __30__

12. 4, 8, 12, 16, __20__

13. 3, 6, 9, 12, __15__

14. 18, 27, 36, 45, __54__

Practice/Homework   PW141

Name _____

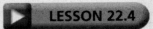 LESSON 22.4

Multiply by 1 and Multiply by 0

Write the product.

1. $0 \times 4 = \underline{0}$	2. $1 \times 5 = \underline{5}$	3. $3 \times 1 = \underline{3}$
4. $7 \times 1 = \underline{7}$	5. $6 \times 0 = \underline{0}$	6. $1 \times 8 = \underline{8}$
7. $3 \times 0 = \underline{0}$	8. $0 \times 9 = \underline{0}$	9. $6 \times 1 = \underline{6}$
10. $1 \times 4 = \underline{4}$	11. $8 \times 1 = \underline{8}$	12. $0 \times 3 = \underline{0}$
13. $0 \times 5 = \underline{0}$	14. $9 \times 1 = \underline{9}$	15. $5 \times 0 = \underline{0}$
16. $1 \times 6 = \underline{6}$	17. $1 \times 7 = \underline{7}$	18. $1 \times 2 = \underline{2}$

 Mixed Review

Add.

19. 465
    + 228

     693

20. 232
    + 226

     458

21. 578
    + 315

     893

22. 765
    + 153

     918

PW142 Practice/Homework

Name _____   ▶ **LESSON 22.5**

Skip-Count to Multiply

How many mittens in all?

Write the answer.

1.

$1 \times 2 = \underline{2}$

2.

$3 \times 2 = \underline{6}$

3.

$5 \times 2 = \underline{10}$

4.

$2 \times 2 = \underline{4}$

5.

$4 \times 2 = \underline{8}$

6.

$6 \times 2 = \underline{12}$

▶ **Mixed Review**

Subtract.

7. 209   8. 456   9. 538   10. 782
   −117   −329   −218   −290
   ‾‾‾‾   ‾‾‾‾   ‾‾‾‾   ‾‾‾‾
    92   127   320   492

Practice/Homework   **PW143**

Name _____

 LESSON 22.6

Problem Solving • Make a Table

Complete the multiplication table.
Start with what you know.
Look for patterns to check your answers.

MULTIPLICATION TABLE

×	0	1	2	3	4	5	6	7	8	9
0	0	0	0	0	0	0	0	0	0	0
1	0	1	2	3	4	5	6	7	8	9
2	0	2	4	6	8	10	12	14	16	18
3	0	3	6	9	12	15	18	21	24	27
4	0	4	8	12	16	20	24	28	32	36
5	0	5	10	15	20	25	30	35	40	45
6	0	6	12	18	24	30	36	42	48	54
7	0	7	14	21	28	35	42	49	56	63
8	0	8	16	24	32	40	48	56	64	72
9	0	9	18	27	36	45	54	63	72	81

Name _____

LESSON 22.7

Use a Multiplication Table

×	0	1	2	3	4	5	6	7	8	9
0	0	0	0	0	0	0	0	0	0	0
1	0	1	2	3	4	5	6	7	8	9
2	0	2	4	6	8	10	12	14	16	18
3	0	3	6	9	12	15	18	21	24	27
4	0	4	8	12	16	20	24	28	32	36
5	0	5	10	15	20	25	30	35	40	45
6	0	6	12	18	24	30	36	42	48	54
7	0	7	14	21	28	35	42	49	56	63
8	0	8	16	24	32	40	48	56	64	72
9	0	9	18	27	36	45	54	63	72	81

Use the multiplication table. Find the product.

1. $3 \times 2 = $ __6__
2. $3 \times 5 = $ __15__
3. $3 \times 8 = $ __24__
4. $4 \times 2 = $ __8__
5. $4 \times 4 = $ __16__
6. $4 \times 6 = $ __24__
7. $6 \times 1 = $ __6__
8. $6 \times 4 = $ __24__
9. $6 \times 7 = $ __42__
10. $8 \times 3 = $ __24__
11. $8 \times 5 = $ __40__
12. $8 \times 8 = $ __64__

▶ **Mixed Review**

Find each sum or difference.

13. $16 + 20 = $ __36__ $78 + 19 = $ __97__ $56 - 28 = $ __28__
14. $64 - 36 = $ __28__ $43 - 17 = $ __26__ $21 + 69 = $ __90__

Practice/Homework PW145

Name _____

 LESSON 23.1

Equal Groups: Size of Groups

Divide into equal groups. Some may be left over.
Draw to show the groups. Write how many. **Check children's work.**

1. Divide 12 apples into 3 equal groups.

__4__ in each group

__0__ left over

2. Divide 9 oranges into 2 equal groups.

__4__ in each group

__1__ left over

3. Divide 16 pears into 3 equal groups.

__5__ in each group

__1__ left over

▶ **Mixed Review**

Count the parts. Write each fraction.
Write the fraction for the whole.

4. 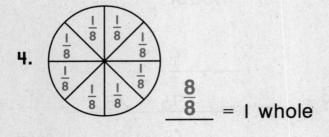 $\frac{8}{8}$ = 1 whole

5. 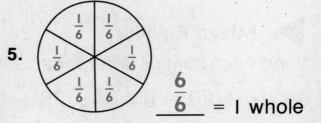 $\frac{6}{6}$ = 1 whole

PW146 Practice/Homework

Name _____

LESSON 23.2

Equal Groups: Number of Groups

Circle equal groups.
Write how many groups there are.
Write how many are left over.

1. Divide 17 ladybugs into groups of 5.

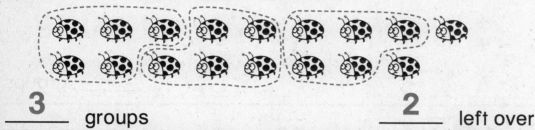

___3___ groups   ___2___ left over

2. Divide 13 ants into groups of 6.

___2___ groups   ___1___ left over

3. Divide 21 beetles into groups of 3.

___7___ groups   ___0___ left over

 Mixed Review

Which measuring tool would you use to measure?
Write **ruler**, **yardstick**, or **thermometer**.

4. the length of a pen   ___ruler___

5. the temperature of a bowl of soup   ___thermometer___

6. the height of a window   ___yardstick___

Practice/Homework   PW147

LESSON 23.3

Name _____

Subtraction and Division

Use ● to show the total. Subtract the number in each group. How many groups of ● can you make?

1. You have 8 ●. Make groups of 4.

   ```
     8       4
   − 4     − 4
   ───     ───
     4       0
   ```
   _____2_____ groups

2. You have 12 ●. Make groups of 2.

   ```
    12     10      8      6      4      2
   − 2    − 2    − 2    − 2    − 2    − 2
   ───    ───    ───    ───    ───    ───
    10      8      6      4      2      0
   ```
   _____6_____ groups

3. You have 15 ●. Make groups of 3.

   ```
    15     12      9      6      3
   − 3    − 3    − 3    − 3    − 3
   ───    ───    ───    ───    ───
    12      9      6      3      0
   ```
   _____5_____ groups

▶ **Mixed Review**

Write = or ≠.

4. 300 + 200 (=) 500

5. 100 + 400 (=) 500

6. 500 + 300 (≠) 700

7. 200 + 600 (=) 800

PW148   Practice/Homework

Name _____

LESSON 23.4

Problem Solving • Act It Out

Use counters to act out the problem.
Make a picture to show your work.

Check children's work.

1. LaToya has 12 marbles. She sorts them into 6 equal groups. How many marbles are in each group?

   __2__ marbles

   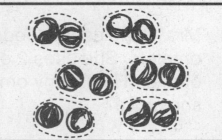

2. Walt has 8 toy cars. He puts them into two equal rows. How many toy cars are in each row?

   __4__ toy cars

3. Janell has 14 plums. She would like to give 2 to each friend. How many friends can she give plums to?

   __7__ friends

4. There are 4 children at Amanda's house. Each child has 3 toys. How many toys are there in all?

   __12__ toys

Create your own story.
Have a friend act it out to solve. Check children's work.

5. Antonio has _____.
   He puts them into _____ groups.
   How many _____ are in each group?

   _____

Practice/Homework PW149

Name _____

LESSON 23.5

Problem Solving • Choose the Computational Method

Choose a method to solve. Show your work. **Check children's work.**

1. Mrs. Potts uses 12 eggs to make omelets. She uses 2 eggs in each omelet. How many omelets does she make?

   __6__ omelets

2. Jason buys a book for $6.03. He buys a bookmark for $0.79. How much money does he spend in all?

   $ __6.82__

3. Tom has 18 photos. He puts 6 photos on each page of his photo album. How many pages does he use?

   __3__ pages

4. There are 846 people at the play. 225 are adults, and the rest are children. How many children are at the play?

   __621__ children

Practice/Homework